Teach Yourself

VISUALLY™

Astronomy

Astronomy

by Richard Talcott

Wiley Publishing, Inc.

Teach Yourself VISUALLY™ Astronomy

Published by Wiley Publishing, Inc., Hoboken, New Jersey

Library of Congress Control Number: 2008938865

ISBN: 978-0-470-34382-1

Printed in the United States of America

10 9 8 7 6 5 4 3 2 1

Book production by Wiley Publishing, Inc. Composition Services

Praise for the Teach Yourself VISUALLY Series

I just had to let you and your company know how great I think your books are. I just purchased my third Visual book (my first two are dog-eared now!) and, once again, your product has surpassed my expectations. The expertise, thought, and effort that go into each book are obvious, and I sincerely appreciate your efforts. Keep up the wonderful work!

—Tracey Moore (Memphis, TN)

I have several books from the Visual series and have always found them to be valuable resources.

—Stephen P. Miller (Ballston Spa, NY)

Thank you for the wonderful books you produce. It wasn't until I was an adult that I discovered how I learn—visually. Although a few publishers out there claim to present the material visually, nothing compares to Visual books. I love the simple layout. Everything is easy to follow. And I understand the material! You really know the way I think and learn. Thanks so much!

—Stacey Han (Avondale, AZ)

Like a lot of other people, I understand things best when I see them visually. Your books really make learning easy and life more fun.

—John T. Frey (Cadillac, MI)

I am an avid fan of your Visual books. If I need to learn anything, I just buy one of your books and learn the topic in no time. Wonders! I have even trained my friends to give me Visual books as gifts.

—Illona Bergstrom (Aventura, FL)

I write to extend my thanks and appreciation for your books. They are clear, easy to follow, and straight to the point. Keep up the good work! I bought several of your books and they are just right! No regrets! I will always buy your books because they are the best.

—Seward Kollie (Dakar, Senegal)

Credits

Acquisitions Editor
Pam Mourouzis

Development Editor
Mike Thomas

Production Editor
Suzanne Snyder

Copy Editor
Marylouise Wiack

Technical Editor
Frank Reddy

Editorial Manager
Christina Stambaugh

Publisher
Cindy Kitchel

Vice President and Executive Publisher
Kathy Nebenhaus

Interior Design
Kathie Rickard
Elizabeth Brooks

Cover Design
José Almaguer

Dedication

I'd like to dedicate this book to my wife, Evelyn, whose everlasting love makes life a joy and whose patience and support helped this book become a reality.

About the Author

Richard Talcott is a senior editor of *Astronomy* magazine, the world's largest magazine devoted to the science and hobby of astronomy. He has written more than one hundred feature articles on both the science of astronomy and observing the night sky.

He edits the "Sky This Month" section at the center of *Astronomy* and creates most of the magazine's star charts.

In collaboration with Joel Harris, he authored *Chasing the Shadow: An Observer's Guide to Eclipses* (Kalmbach Publishing Co., 1994). He has been an avid observer of the night sky since the 1960s and has witnessed seven total solar eclipses.

He graduated from Marietta College in Marietta, Ohio, in 1976 with a degree in Mathematics. After attending graduate school in astronomy at Ohio State University, he returned to Marietta in the early 1980s as a Lecturer in the Physics Department. Since 1986, he has worked as an editor at *Astronomy.* He lives in Waukesha, Wisconsin, with his wife, Evelyn.

Acknowledgments

Although my name appears alone on this book's cover, I can claim only partial credit for what you now hold in your hands. I'd first like to thank Marilyn Allen and Bill Liberis for bringing me to this project. But the true stars of the production are the people at Wiley who saw this book through from concept to completion. Special thanks go to Pam Mourouzis, Mike Thomas, and Suzanne Snyder for their efforts to keep me on time and in line.

The words in an astronomy book typically play second fiddle to the photographs accompanying them. This one is no different. Although the beauty of the universe is there for all to see, capturing its splendor takes hard work, dedication, and talent. All the photographers represented here possess these attributes in abundance. Their abilities are exceeded only by their generosity in letting me share their images with you. My thanks go to (in alphabetical order): Anthony Ayiomamitis, Bill and Sally Fletcher, Robert Gendler, Alister Ling, Marvin Nauman, Jack Newton, Martin Ratcliffe, Rob Ratkowski, Fred Ringwald, Mike Salway, Mike Simmons, Luca Vanzella, and Oshin Zakarian. Most of the other great images came from NASA spacecraft — including the Hubble Space Telescope, the Solar and Heliospheric Observatory, and various missions to the planets. Thanks also go to the Lunar and Planetary Institute for the close-up images of the Moon, and to the National Optical Astronomy Observatories and their Advanced Observing Program at Kitt Peak.

My friends and colleagues at *Astronomy* magazine also played a major role in bringing this book to fruition. Editor Dave Eicher and senior editors Michael Bakich and Frank Reddy contributed images, sketches, and valuable feedback. Artist Roen Kelly produced many of the illustrations that render sometimes difficult concepts understandable. It's a pleasure working with each and every one of them. A special thank you to Frank Reddy, who served as this book's technical editor. Any mistakes that made it to print are my responsibility, of course, but Frank's keen eye kept these to a minimum.

Table of Contents

Introduction to the Sky

Choose the Right Equipment

chapter 3 View the Naked-Eye Sky

chapter 4 Explore the Winter Sky

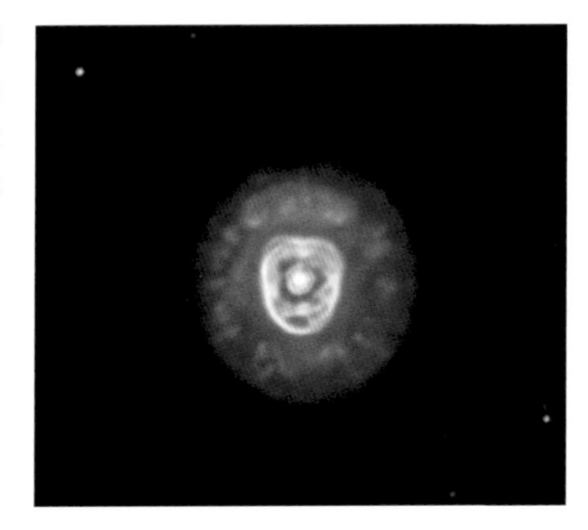

Explore the Spring Sky

Explore the Summer Sky

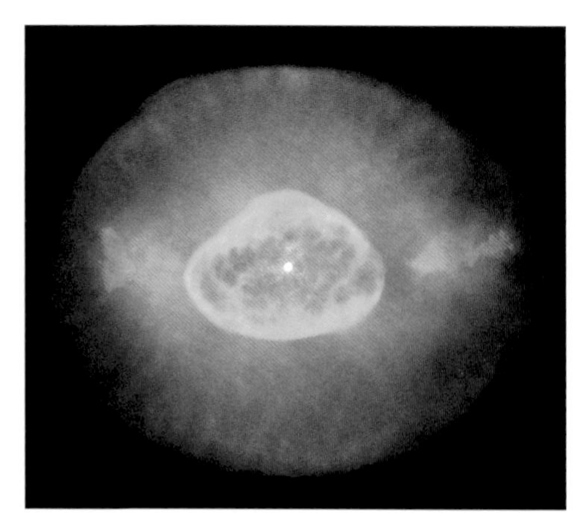

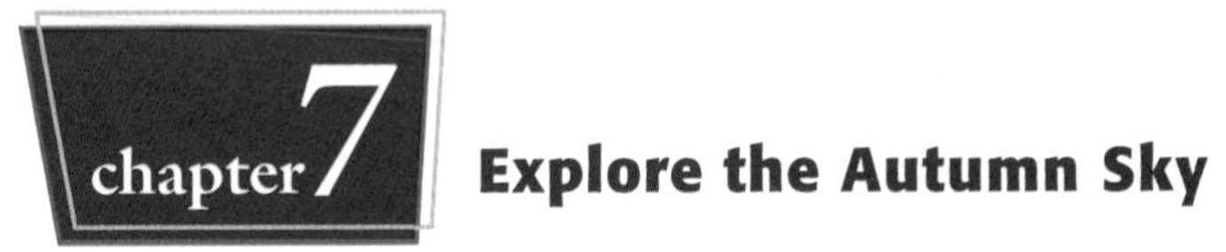

chapter 7 Explore the Autumn Sky

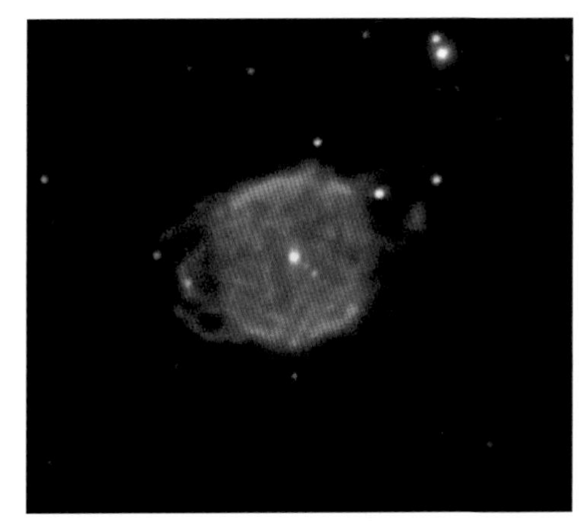

chapter 8 Observe the Sun and Moon

chapter 9 Observe the Rest of the Solar System

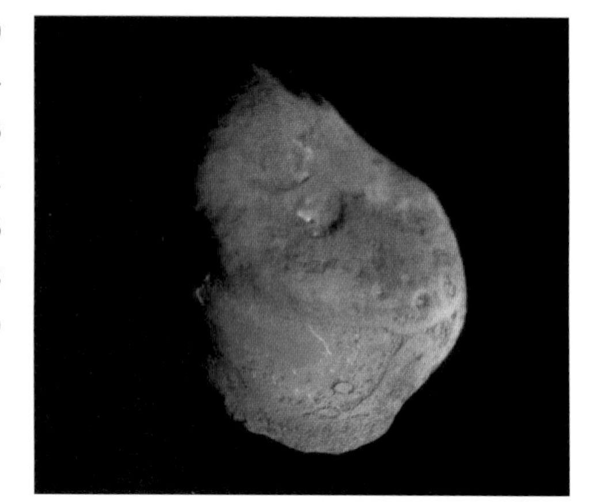

chapter 10 Observe the Deep Sky

Record the Sky

chapter 

Introduction to the Sky

On a dark night well away from the city, the sky seems to explode with stars. To help orient themselves, ancient skywatchers named the brightest objects—names we still use today. In more modern times, astronomers have cataloged thousands of stars and deep-sky objects, and fixed their positions on a celestial grid.

Scientists have also learned the sizes and distances of most of the objects we observe. Of course, all this knowledge does observers little good if the glow from city lights hides the sky from view. Finding a good viewing site can take some effort, but it's worth it. You'll often find some of the best observing at "star parties" attended by fellow skygazers.

Star Names

What's that bright star over there? If you spend much time observing the night sky, you're bound to hear this question over and over. Once you become familiar with the starry canopy, the answer will come to you almost immediately. "Why, that ruddy star hanging low in the south is Antares, and that gleaming white star overhead is called Vega." Most star names commonly used today came from the Arabic language, although many of these names originated with the ancient Greeks.

Common star names abound throughout the heavens, although most remain far from household names. Orion the Hunter has many outstanding traits, not least among them the fact that all of its bright stars have well-known monikers. Betelgeuse, which marks Orion's upper-left corner, ranks high on most people's list of favorite star names. It appears to be a corruption of the Arabic *yad al-jawza*, meaning "hand of al-jawza." (*Al-jawza* itself means either "the giant" or "the central one.")

In Orion's opposite corner lies the constellation's other luminary, Rigel. Its name derives from the Arabic *rijl al-jawza*, meaning "foot of al-jawza." At the lower left of the Hunter's figure sits Saiph. This star's name has a strange history. It comes from a longer Arabic phrase meaning "sword of the giant." But Saiph has no relation to Orion's sword—a string of stars that descends from Orion's belt—apparently the name got transferred by mistake long ago, and it stuck. Rounding out Orion's four corners is Bellatrix. Unlike its Arabic brethren, Bellatrix is a Latin name that means "Amazon Star."

That leaves us with Orion's three belt stars. At the western (right) end lies Mintaka and at the eastern end Alnitak. Both come from Arabic phrases meaning the "belt of al-jawza." Alnilam, the belt's central star, also gets its name from Arabic. The name means "string of pearls," an apt designation for the three-star belt.

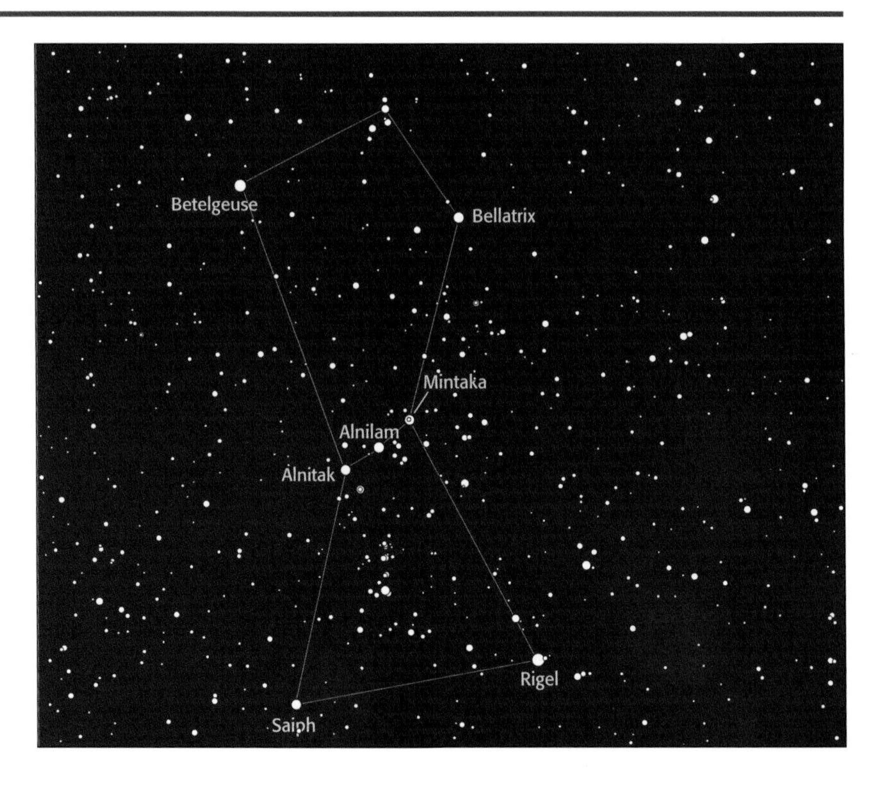

TIP

Directions in the sky can be confusing. Star maps typically show north up and east to the left because that's the way we see the sky. Imagine viewing a constellation that lies due south—the northern part of the pattern then appears at top, but the eastern side faces left.

Astronomical Catalogs

Astronomers would be in dire straits if they had to come up with a common name for every star and deep-sky object out there. Smartly, they've devised several classification schemes to tackle the problem. Many of the brighter objects have earned multiple designations, with the first bestowed usually the most common. But you can be sure about one thing: Nearly every object in the sky—even newly discovered ones—has a designation attached to it.

You might think that with perfectly good names like Betelgeuse and Rigel, Orion's brightest stars wouldn't need any other titles. That's not the way astronomers think. In their never-ending quest to catalog objects, scientists have developed more complete and systematic listings, and it really wouldn't work to leave out the brightest objects.

In 1603, German lawyer Johannes Bayer published a star atlas called *Uranometria*. In it, he designated the brightest stars in each constellation by a Greek letter. In most cases, the letters went in order of decreasing brightness, so the brightest star was Alpha (α), the second Beta (β), and so on. In Orion, Betelgeuse received the Alpha designation and Rigel is Beta. In the eighteenth century, a catalog published by British astronomer John Flamsteed numbered the brighter stars in each constellation starting at the western edge and working eastward. So, Betelgeuse is also 58 Orionis and Rigel is 19 Orionis. Still later catalogs, used mostly by professionals, also incorporate Orion's luminaries.

Deep-sky objects are no different. In 1784, French astronomer Charles Messier compiled the first such catalog. The famed Orion Nebula earned the designation M42. In the nineteenth century, Danish astronomer John Dreyer compiled the *New General Catalogue*, a compendium of several thousand objects (Messier's barely surpasses 100). The Orion Nebula is the 1,976th object in this catalog.

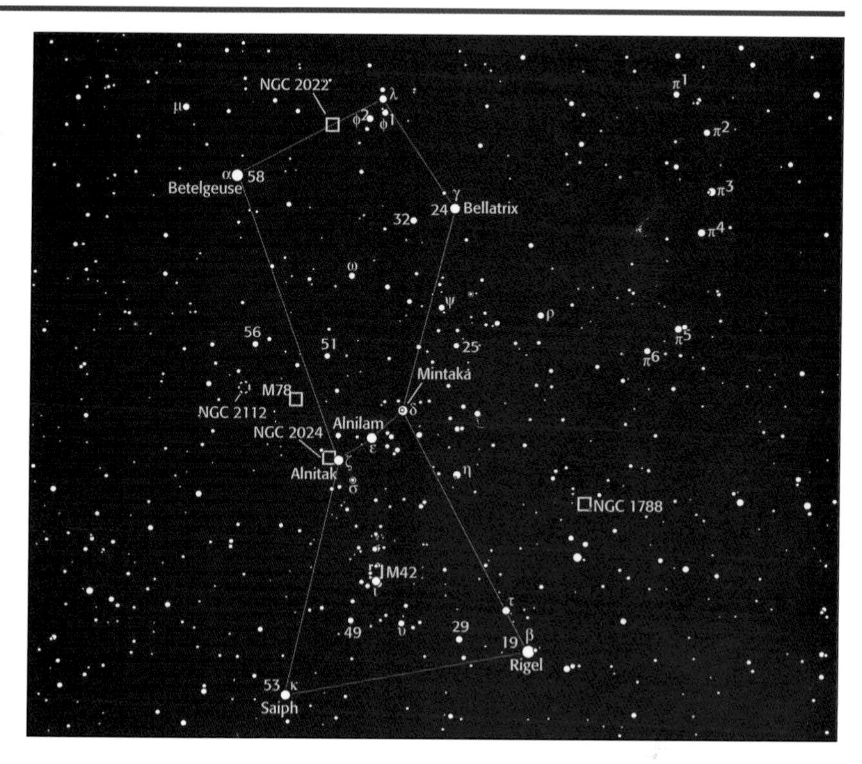

Understanding Magnitudes

Few subjects confuse beginners more than the magnitude system astronomers use to describe the brightnesses of celestial objects. The problem is that the system is counter-intuitive, with bigger numbers corresponding to fainter objects. Once you know how the system is set up, however, you shouldn't have any trouble.

The story of magnitudes dates to the Greek astronomer Hipparchus. In the 2nd century B.C.E., he cataloged the stars he could see from his home. He called the brightest stars *1st magnitude,* the next group *2nd magnitude,* and so on to the faintest stars at *6th magnitude.* Later astronomers quantified this magnitude system, keeping lower numbers for brighter objects and extending it into negative magnitudes for the very brightest.

Magnitudes of Celestial Objects

Magnitude	***Object or Limit***
−26.7	Sun
−12	Full Moon
−4.5	Venus at its brightest
−3.9	Venus at its faintest
−2.9	Mars at its brightest
−1.47	Sirius (the brightest star)
−0.72	Canopus
−0.29	Alpha Centauri
−0.04	Arcturus
0.03	Vega
6.5	Naked-eye limit
9	Binoculars limit
13	8-inch telescope limit
18	Approximate large scope limit
30	Hubble Space Telescope limit

In quantifying the magnitude system of Hipparchus, astronomers realized the Greek scientist's 1st-magnitude stars were about 100 times brighter than his 6th-magnitude stars. So, they devised a system with a logarithmic scale wherein a 5-magnitude difference would equal a factor of 100 in brightness. This table gives brightness ratios for several magnitude differences.

The Magnitude System

Magnitude Difference	***Brightness Ratio***
0.5	1.58
1.0	2.51
1.5	3.98
2.0	6.31
2.5	10.0
3.0	15.8
4.0	39.8
5.0	100
10.0	10,000
15.0	1,000,000

Seeing Color in the Sky

To see color, the eye needs a fair amount of light to reach the retina. Bright stars and auroras are up to the task for naked-eye observers. Use the light-gathering power of binoculars or a telescope, and you'll see colors in many more stars. Although the beautiful reds and blues you see in photographs of deep-sky objects are real, an observer staring through an eyepiece has no hope of seeing them.

Color shows up best in bright stars. Leo's 1st-magnitude luminary, Regulus, sports a distinct bluish hue. This reflects the star's high surface temperature. (In astronomy, blue is a hot color and red is a cool color.)

Oddly enough, the brightest star in the sky shows no natural color. Sirius has just the right surface temperature so that it emits equal amounts of light at all wavelengths—what we perceive as white. When Sirius lies low in the winter sky, however, it often looks colorful. The rapidly changing colors appear because the star's light passes through our turbulent atmosphere. This phenomenon is called "twinkling."

Colorful shades of red also show up in stars. With 1st-magnitude Antares (near right), the color tends toward orange, reflecting the star's modest surface temperature. But in the comparatively cool atmosphere of R Leporis (far right), almost all the light radiated lies at the red end of the spectrum. Colors like this show up in the eyepiece when you have a large enough scope operating at a fairly high magnification.

Dark Adaptation

The human eye is a phenomenal organ, but it does have its limitations. Walk from bright sunshine into a darkened room, and it takes a few minutes before your eyes can adjust. Similarly, if you're in a brightly lit room at night and then step outside, you won't see clearly for quite a while. To get the most out of an observing session, you need to be dark adapted. It's easy to do, and will make a world of difference in what you see.

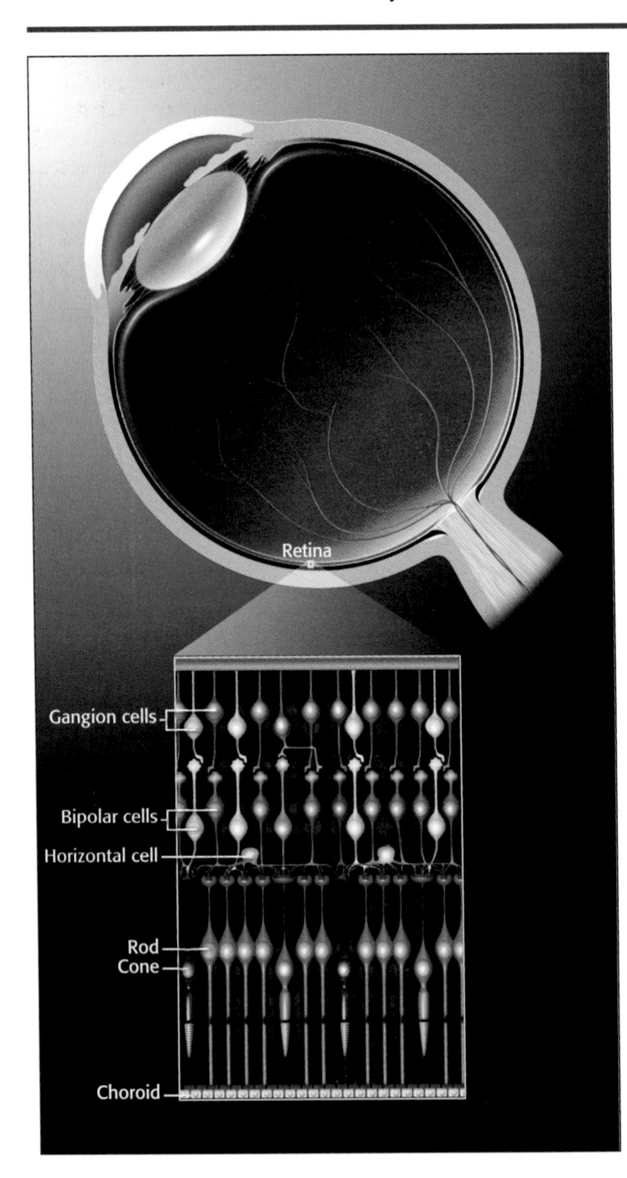

Go from bright light to dark, and your eye will do two things. First, your pupils will dilate to let in more light. In daylight, the pupil may open only 1 to 2 millimeters, but that can jump to 5 to 7 millimeters in darkness. Second, the eye's rod cells produce more rhodopsin, which increases their light sensitivity. It can take up to 30 minutes before the eye fully dark adapts and can see the most detail. When observing, avoid bright lights if at all possible.

The eye's rod cells are affected less by red light than by blue light. So, if you need some illumination to see telescope or camera controls, or to avoid objects on the ground, use a red flashlight. (You can either buy one or make one by taping red cellophane over the flashlight's front.) If you can, adjust the flashlight's light level so it illuminates just what you need to see.

Averted versus Direct Vision

It sounds odd to most beginners, but it's true: You can see fainter objects by not looking right at them. If you want to see detail in an object, then certainly look directly at it. But if you want to detect a dim object that seems just out of reach, avert your gaze slightly. You'll be amazed at what pops into view.

If you want to see detail along the Moon's terminator, or in the cloud tops of Jupiter's atmosphere, you need to let the light fall on the eye's central *fovea*. This area is lined with day-sensitive cone cells, which let you see fine details in bright lighting conditions. But many of the sights observers want to target are dim, and averted vision can add half a magnitude to what you see. When using a telescope, try to look about halfway from the center of the field (presumably where you placed the object) to the edge. Averted vision will also deliver the clearest view of stars in close proximity, such as Ursa Major's Mizar and Alcor.

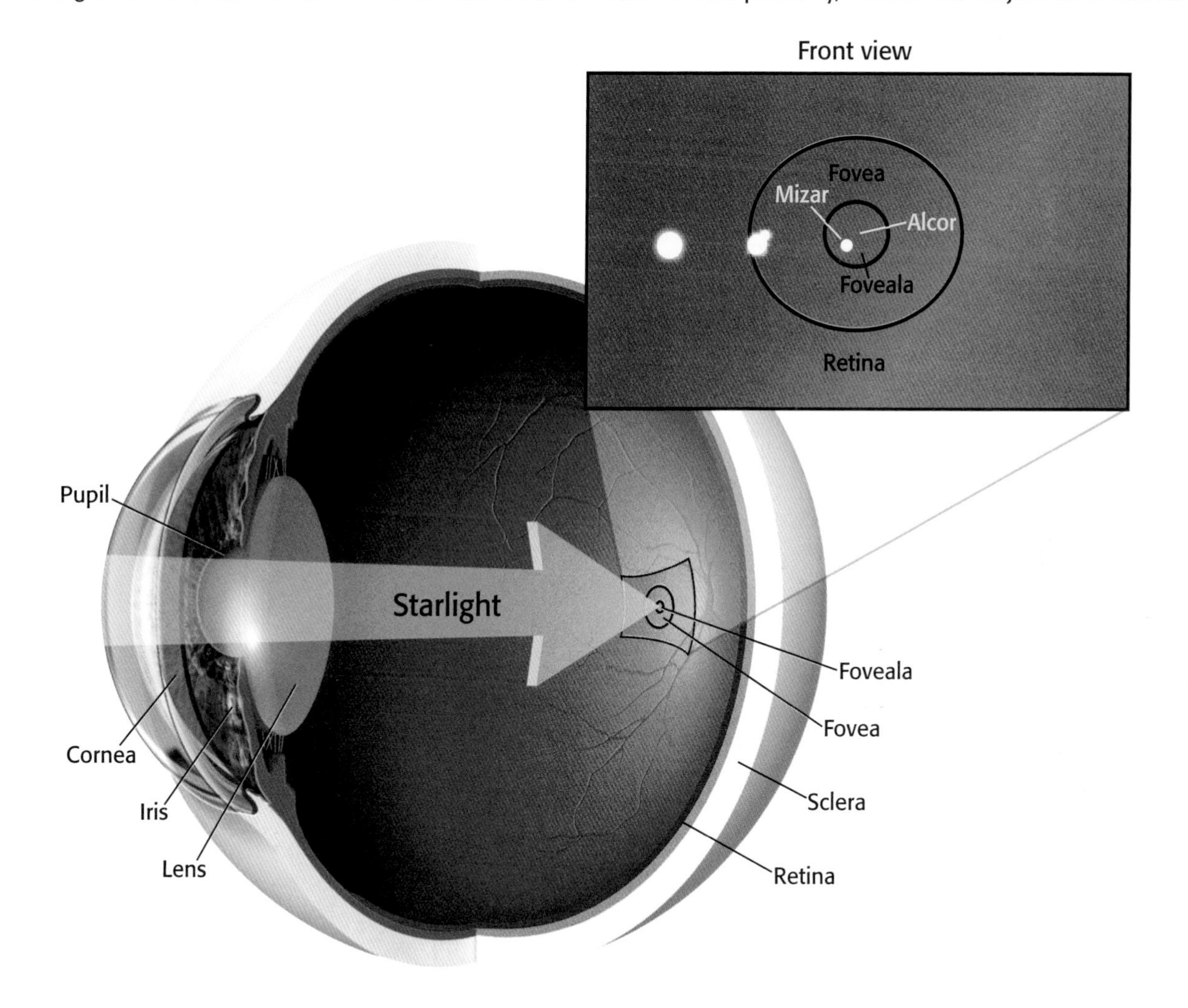

Finding Your Way around the Sky

If you want to pinpoint a place on Earth, all you need to know is its latitude and longitude. The same technique works in the sky, where astronomers have established similar coordinate systems. Once you know where an object is, you can center it in your telescope either by moving it by hand or by having a computer slew it for you. You'll find that observers hold strong views about which is the better method.

The most intuitive approach to finding your way around the sky is by using the *horizon coordinate system*. Here, you measure how far above the horizon an object lies (its *altitude*) and how far east of due north it lies (its *azimuth*). More often than not, you'll read or hear directions given in this system. For example: "Venus lies 15° high in the west an hour after sunset."

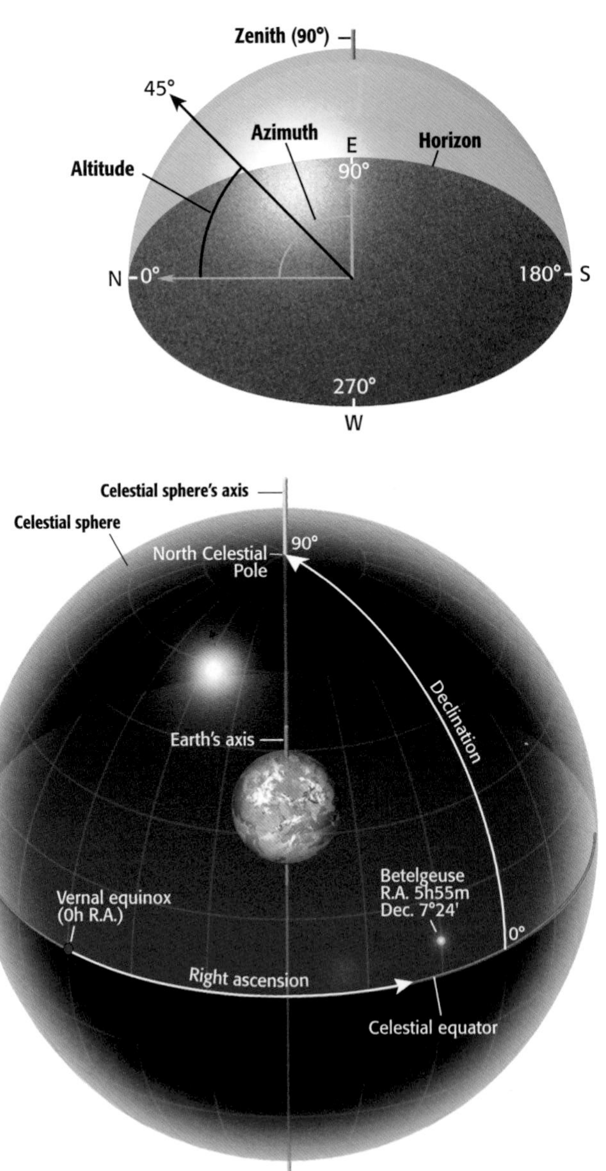

The *equatorial coordinate system* closely matches the latitude and longitude we use on Earth. Imagine all celestial objects lying on the surface of an infinitely large celestial sphere centered on Earth. The celestial equator is an extension of Earth's equator into the sky, and the celestial poles mark where Earth's axis of rotation intersects the celestial sphere.

Astronomers measure how far north or south of the celestial equator an object lies (its *declination*) and how far east of the vernal equinox an object lies (its *right ascension*). The advantage of the equatorial coordinate system is that it remains essentially fixed relative to the stars. So, if you know the right ascension and declination of the star Betelgeuse tonight, it will be in the same position next week, next year, and even next decade. In the horizon system, an object doesn't stay in the same place from one hour to the next.

Our winter friend Orion stands astride the celestial equator (the line of 0° declination). Betelgeuse lies nearly 10° of declination north of the equator and Rigel appears about the same distance south. The main body of Orion falls between 5 hours and 6 hours of right ascension, placing it about one-quarter of the way around the sky from the vernal equinox.

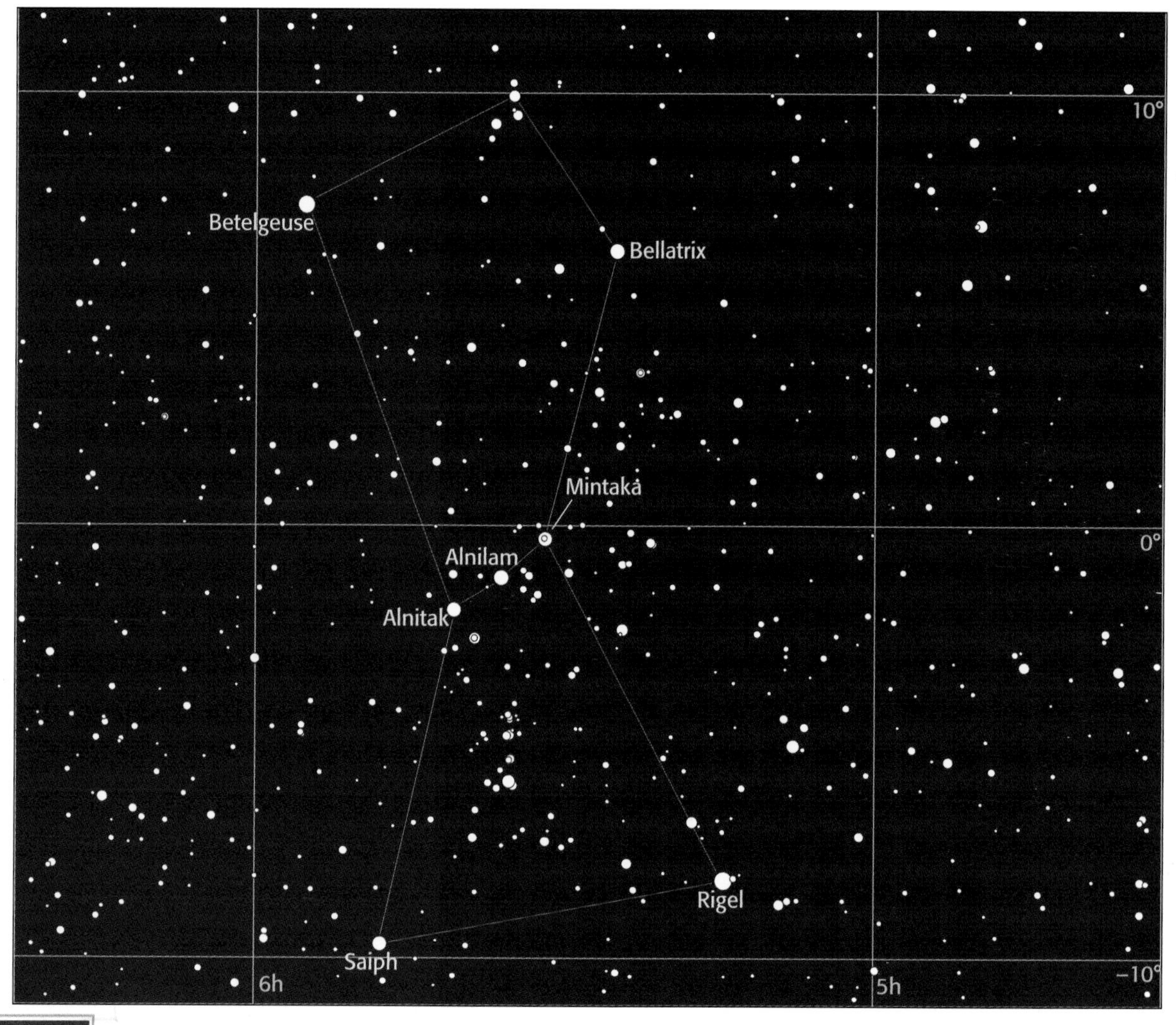

TIP

To Star Hop or Go-to

How do you point your telescope to the Orion Nebula? If you're a traditionalist, you first find Orion's belt, and then follow a line of stars southward until you reach the glowing gas cloud. These people claim you won't learn the sky well until you learn to hop from star to star. If you're a technology freak, you tell your telescope's computer to find M42, and it slews there automatically. These people claim they see far more objects through their go-to telescopes because they don't "waste" time searching. Both groups are right, at least to some extent.

Apparent Sizes and Distances

It's fairly easy to get confused about the sizes of celestial objects. The Moon looks bigger near the horizon than when it's up higher. But many people think the Moon is nearly as large as the Big Dipper, and must certainly span a greater diameter than the Pleiades star cluster. With a few handy methods to measure the sizes of sky objects, you won't be making these mistakes.

If you hold a finger at arm's length, it will span an angle of approximately 2°. That's more than enough to totally block the Moon, but it's not enough to completely cover the Andromeda Galaxy. And most constellations extend much farther than your finger could hope to block.

If you make a fist and hold it at arm's length, you'll be covering approximately 10° of sky. This handy reference matches the size of the Big Dipper's bowl, but it's smaller than the Great Square of Pegasus.

Open your hand wide and hold it at arm's length, and you're now spanning some 25° from the tip of your thumb to the tip of your pinkie. This larger unit makes a good way to gauge the separation of objects in the sky, say the distances between the stars of the Summer Triangle.

The Sun spans a whopping 865,000 miles and lies an average of 93 million miles from Earth. If you divide the Sun's diameter by its distance, you get a measure of how big it appears in the sky. The result: about 0.5°, or 30 arcminutes, across. All three photographs on this page are shown to the same scale.

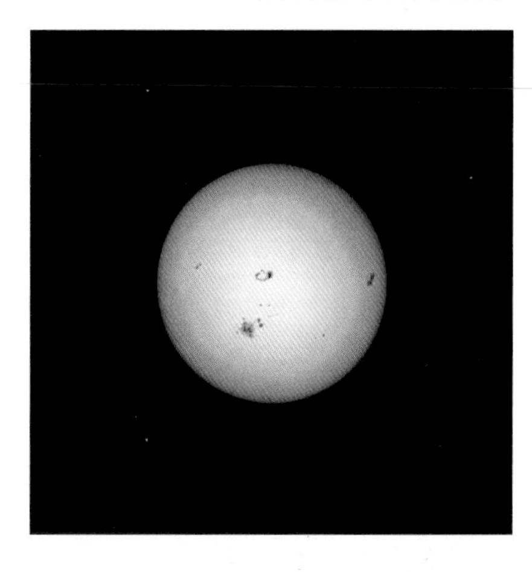

Our Moon is much smaller than the Sun (only 2,160 miles across) but also much closer to Earth (239,000 miles away). Combine these two numbers, and you'll find that the Moon spans the same 0.5° angle as the Sun does. The uncanny coincidence leads to exquisite solar eclipses over a narrow path on Earth's surface.

The Pleiades star cluster lies some 440 light-years from Earth, but its stars spread out across nearly 10 light-years. This makes the Pleiades appear significantly larger in our sky than either the Sun or Moon.

Astronomical Distances

To those of us used to hopping in the car to go grocery shopping, distances in the universe seem astounding. The nearest celestial object, the Moon, lies 239,000 miles away. But that's just the tip of the iceberg. In the solar system, distances climb to millions and billions of miles, and the nearest star system beyond the Sun lies many trillions of miles away. It's no wonder even nearby stars appear as mere points of light.

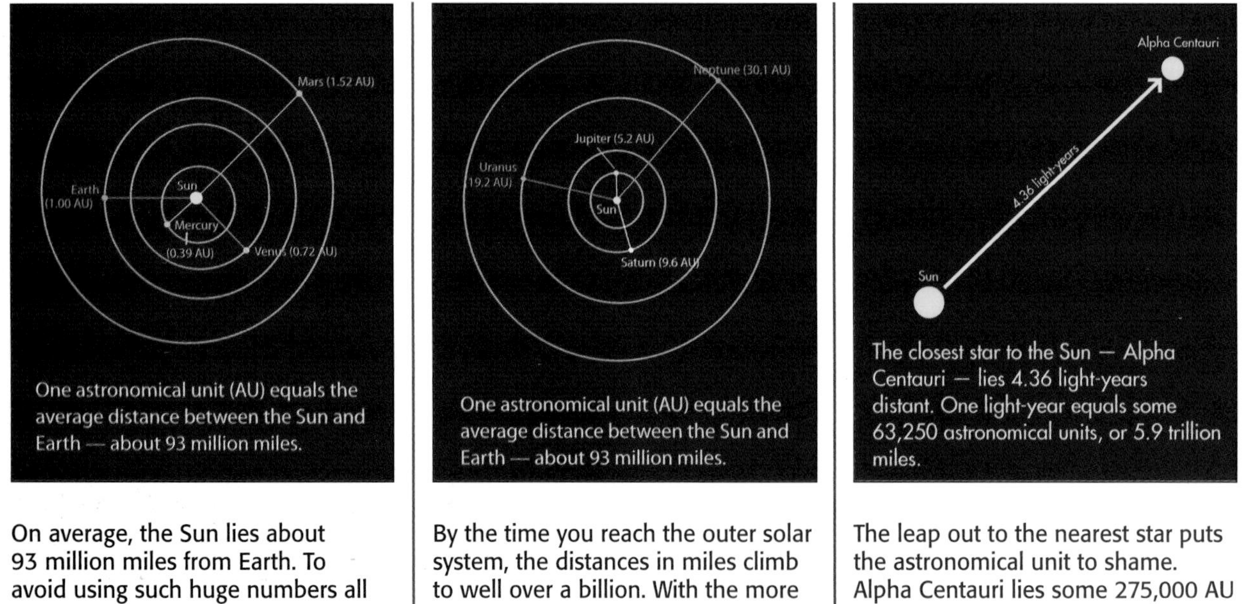

One astronomical unit (AU) equals the average distance between the Sun and Earth — about 93 million miles.

One astronomical unit (AU) equals the average distance between the Sun and Earth — about 93 million miles.

The closest star to the Sun — Alpha Centauri — lies 4.36 light-years distant. One light-year equals some 63,250 astronomical units, or 5.9 trillion miles.

On average, the Sun lies about 93 million miles from Earth. To avoid using such huge numbers all the time, astronomers defined a new measuring stick called the *astronomical unit* (AU), which is equal to the Sun–Earth distance.

By the time you reach the outer solar system, the distances in miles climb to well over a billion. With the more appropriate AU, however, the numbers stay far more reasonable.

The leap out to the nearest star puts the astronomical unit to shame. Alpha Centauri lies some 275,000 AU from Earth, or 25 trillion miles. For such vast distances, astronomers turn to the *light-year*, the distance light (moving at a speed of 186,000 miles per second) travels in a year. Still, Alpha Centauri lies 4.36 light-years away.

Laws that make commerce run more smoothly push us away from the natural rhythms that form the basis of timekeeping. Instead of defining noon as the moment when the Sun lies due south in the sky, our clocks tell us noon happens at the same time in cities hundreds of miles to the east or west. And during summer, the discrepancy grows larger. All this makes it harder to figure out when a celestial event will occur.

When you read about an event happening in the early evening sky, the time will usually be quoted as being half an hour or an hour after sunset. If you want to know the clock time of this event, you'll need to look up when your local sunset occurs. There's simply no way to be precise because sunset occurs at different times in different locales. The good news: In most cases, the event will look the same regardless of where you live.

July 1, 2010, 1 hour after sunset
Looking west

Timing is precise when it comes to eclipses of the Sun and Moon. The alignments that produce these events happen at certain times, and all you need to do to calculate when the eclipse begins, for example, is to correct for your time zone. The phases of the Moon operate in the same way.

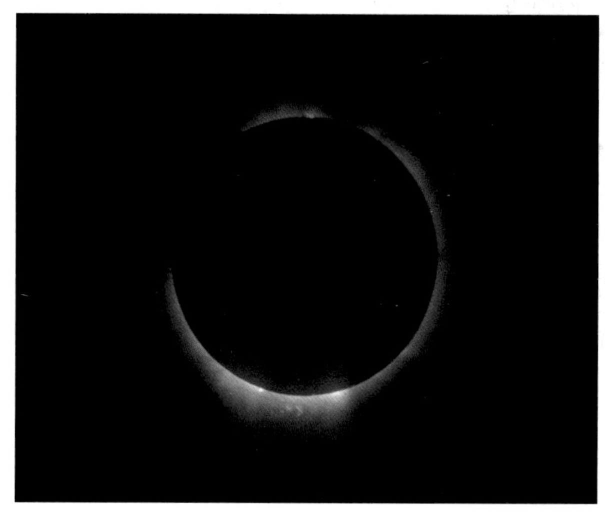

TIP

Universal Time

To get around some of the problems associated with time zones, astronomers developed an unambiguous time standard known as *Universal Time* (UT). Observers anywhere in the world can coordinate their observations by using UT. UT runs five hours ahead of Eastern Standard Time, for example, and 4 hours ahead of Eastern Daylight Time.

How Dark Is Your Sky?

All other things being equal, the darker your sky, the more you can see. But the number of truly dark places in the world is dwindling as population increases. In large parts of North America and Europe, you have to travel hundreds of miles to escape the pall of urban skyglow. How can you tell if you've found a good site? Spend a few minutes calculating the limiting magnitude.

Europe blazes with civilization's lights. The problem: A lot of this light goes up into the sky, where it harms our view of the night sky, and not toward the ground, where it would do the most good. Europeans in search of dark skies usually have a long way to travel.

North America spreads out a lot more than Europe and has a lower population density, so it's no wonder some good observing remains. Still, the lights seemingly keep expanding, except in some parts of the western American deserts and mountains, and in northern Canada. A close look toward the eastern seaboard reveals only a few oases of darkness.

You can use the region of the Big and Little Dippers to determine the darkness of your observing site. Under a dark sky and good observing conditions, an average person can see stars down to about magnitude 6.5. Exceptionally sharp-eyed observers under outstanding conditions can get down to magnitude 7.0 or a bit fainter.

Under the less-than-ideal conditions we more typically experience, the numbers won't be so high. A good way to judge is to look toward the Dippers. The Big Dipper, of course, contains a lot of bright stars, and you should see them easily unless the Moon's shining brightly or haze abounds. The Little Dipper offers a better test. On this photograph, we've marked the magnitudes of several stars (the decimal points have been dropped to avoid confusion with stars). When you're under the sky at night, establish which is the faintest star you can see. That is your *limiting magnitude.*

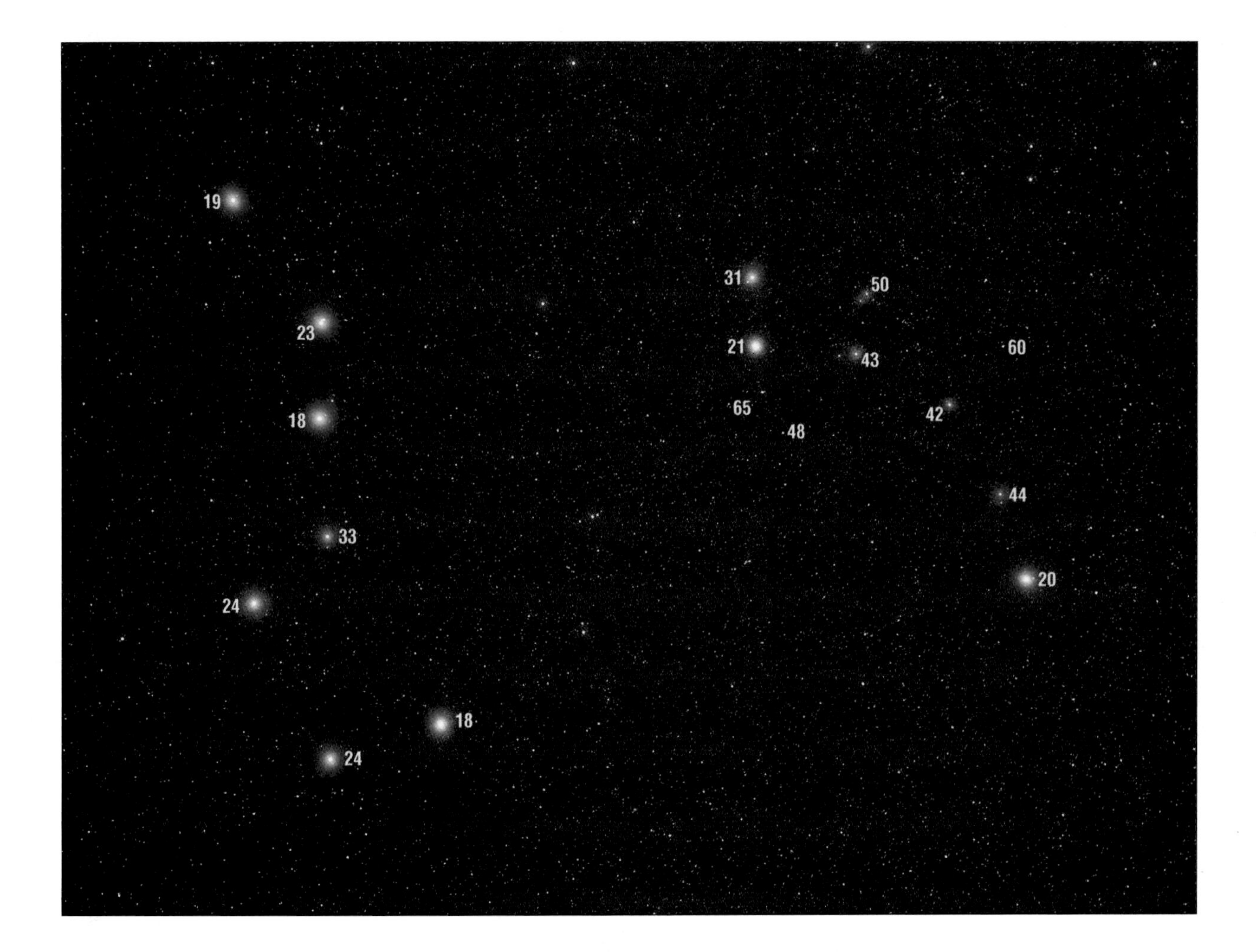

Party under the Stars

To some, observing the night sky is a solitary pursuit. But most people enjoy the camaraderie of their fellow enthusiasts. Some join together to witness the spectacle of a solar eclipse. Others simply gather at a good dark-sky site for a few nights under the stars. And don't think such star parties are only the domain of experts. Beginners can learn a lot about the sky—and about the telescope they might want to purchase—by attending a star party.

Few events draw people together more than a total solar eclipse. Perhaps it's the exotic locales, and people prefer company when they explore uncharted regions. Or maybe it's the expense and hassle of trying to book an entire tour by yourself. But there's no question that a group of people bring excitement to eclipse viewing. You'll never forget the whoops and hollers that come from people as the Sun winks out and totality commences.

Aurora watching also gains by having other people around. During a bright display, things happen quickly, and having several sets of eyes on the sky almost guarantees you won't miss anything. And if you're photographing an aurora, it's easy to get caught up in the picture taking and not take in the grand sweep.

The South Pacific Star Party features dark skies and friendly people—a grand combination. In this long-exposure shot, big telescopes stand watch as stars trail in the distance. The red lights belong to observers protecting their night vision. Most beginners benefit by going to a local star party, whether it's at a dark-sky site or a planetarium or science center. You can pick up a lot of helpful hints, see what the universe looks like through a big scope, and perhaps even test-drive a telescope you're thinking about buying.

chapter 2

Choose the Right Equipment

"What kind of telescope should I get?" That's almost always the first question people ask when they discover an interest in astronomy and observing. Unfortunately, there's no simple answer. It's not even essential that you buy a telescope. One of the night sky's great secrets is that you can devote months or even years to the hobby without owning a telescope.

So, before you put down your hard-earned cash, consider whether binoculars might do the trick. You may be pleasantly surprised at how much binoculars reveal, and they also do a good job of introducing a child to the sky's majesty. If you do opt for a telescope, there's a huge variety on the market. You should be able to find an instrument that will deliver years of observing pleasure.

Naked Eye, Binoculars, or a Telescope?

The challenge facing many beginners is deciding what equipment they need. The answer may be none. After all, the naked eye provides great views of many astronomical subjects. But most people will want at least a little optical aid from time to time. Binoculars will fill that bill perfectly. Then, when you're ready to step up to a telescope, you'll already know what types of objects you like best. And that likely will influence your ultimate choice.

Legions of backyard astronomers got their start by witnessing a particularly impressive meteor shower. Others can trace their love of the night sky back to a spectacular auroral display. Yet no one would even consider getting out a telescope to watch this kind of event. Naked-eye astronomy delivers the sky's broad brush strokes. Constellations and asterisms come to the fore, along with those amazing events simply too big to be contained within a narrow field of view.

Binoculars have three great assets. First, they're a piece of optical equipment a lot of people already own. Second, many people find it more comfortable to observe with two eyes instead of the one normally used with a telescope. And third, the extra light-gathering power brings in many more objects than the naked eye can see. Binoculars give you exquisite views of Milky Way star clouds, large open star clusters, bright comets, and total lunar and solar eclipses.

A small telescope brings many more deep-sky objects into view than binoculars do, and shows much finer detail in solar system objects. If you want to see the rings of Saturn, you'll need to look through a telescope. The same holds true for atmospheric detail on Jupiter and surface detail on the Moon and Mars. A small telescope will also give pleasing views of many star clusters, nebulas, and galaxies.

Consider binoculars the instrument of choice if you're on a budget, or if you have a child interested in the sky but one who seemingly changes his or her mind on a monthly basis. Binoculars have so many uses that you can be sure they won't end up gathering dust. Binoculars also produce right-side-up images. This makes them perfect for terrestrial viewing, but also makes them ideal for finding bright deep-sky objects.

An observer using these 7×21 binoculars won't be searching for faint objects, but instead will be seeking detail in bright objects. In the specs of any binoculars, the first number refers to the magnification and the second to the diameter in millimeters of the front lenses. So, 7×21s deliver a power of 7× but gather light through only a 21mm aperture—only three to four times wider than the dark-adapted human eye.

With these 10×50 binoculars, a person can begin observing the night sky in detail. The 50mm lenses gather about 100 times more light than a middle-aged person's dark-adapted eye. This means that many more objects will come into view. And the 10× power will deliver detailed views of the brightest deep-sky objects.

Once you start getting into large binoculars, typically those with lenses at least 63mm in diameter, you're going to want to mount them on a tripod. It simply becomes too tiring to hold such a heavy instrument for long. This pair of 18×50 binoculars is image stabilized, so you can observe at high power without worrying about image shake. You'll typically pay more than $1,000 for the privilege of using binoculars like this, however.

Types of Telescopes

Once you move beyond binoculars, you've entered the realm of the telescope. A telescope lets you see detail in many objects, from our next-door neighbors in the solar system to distant galaxies. Today's telescope market offers plenty of choices. Always remember that the best telescope is the one you'll use most often. Don't get one that's so unwieldy you'll never want to take it outdoors. A well-used small telescope will give you much more pleasure.

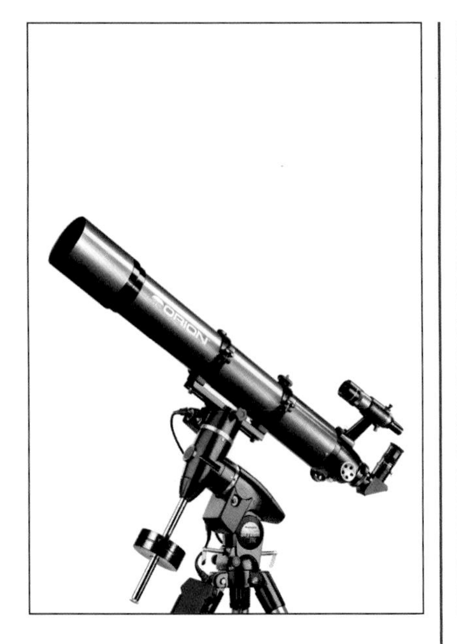

A *refracting telescope* uses a lens—or, more often, a series of lenses—to create an image. Light passing through a lens gets *refracted,* or bent, and brought to a focus. However, a single lens brings each different color to a different focus, a defect called *chromatic aberration*. Achromatic and apochromatic refractors reduce these problems. Although refractors usually deliver superb images, they cost more per inch of aperture than other models.

A *reflecting telescope* uses a mirror to bring light to a focus. Because no light passes through the glass, there's no chromatic aberration. It's also easier to manufacture a reflector's single glass surface than the multiple surfaces in a refractor. So, big reflectors are cheaper than modest refractors. However, because light bounces back up a reflector's tube, a second mirror has to deflect light out of the main path. This obstruction can reduce image quality.

A *compound,* or *catadioptric, telescope* uses elements of both reflectors and refractors. Typically, a corrector plate lies at the front of the telescope. Light refracts through the plate before being sent on to the primary mirror, which creates the image. The big advantage of a compound telescope is that the light path is folded, so the resulting telescope tends to be smaller and more portable.

Trip the Light Fantastic

The passage of light through a telescope may seem complicated, but it follows the laws of physics. The laws of refraction and reflection simply tell the light how to behave.

White light that passes through a simple lens gets broken into its component colors, just as if it were passing through a prism. A compound lens largely corrects this aberration. The focal length of any telescope equals the distance from the main lens or mirror to the point where light comes to a focus.

Refracting telescope

Objective

Eyepiece

Light traveling through a reflecting telescope strikes a spherical or parabolic primary mirror and then bounces back in the same direction it came from. A secondary mirror then redirects the light out of the path. In a Newtonian design, the light bounces out the side of the tube. In a Cassegrain design, the light bounces back toward the main mirror and through a hole in that mirror.

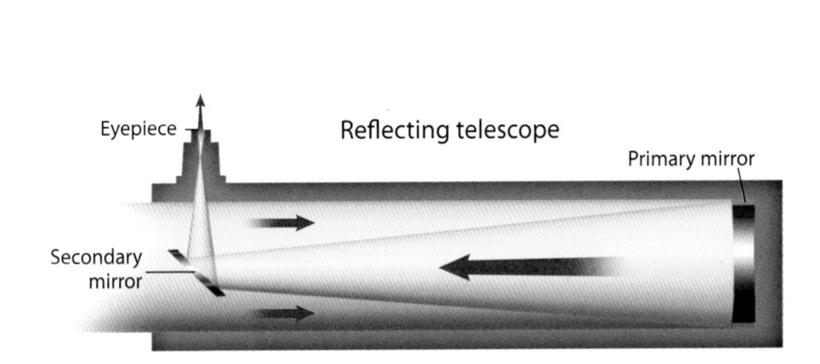

Light's trip through a compound telescope is much more complicated. Here, the light passes through a corrector plate, which bends it toward the primary mirror. The mirror then bounces it back up the tube, where a secondary mirror handles it the same way a normal reflecting telescope of Cassegrain design does.

Catadioptric telescope

Telescope Mounts

Most people think the optics are the most important feature of a telescope. But equally valuable is having a sturdy mount. A wobbly mount will deliver shaky images—and frustration in spades. Mounts come in two basic flavors: alt-azimuth and equatorial. Either one can be equipped with go-to technology that enables the telescope to slew immediately to any object in the scope's computer database.

An *alt-azimuth mount* lets you swing the telescope along two axes. One direction moves the telescope vertically and the other moves it horizontally, paralleling the horizon.

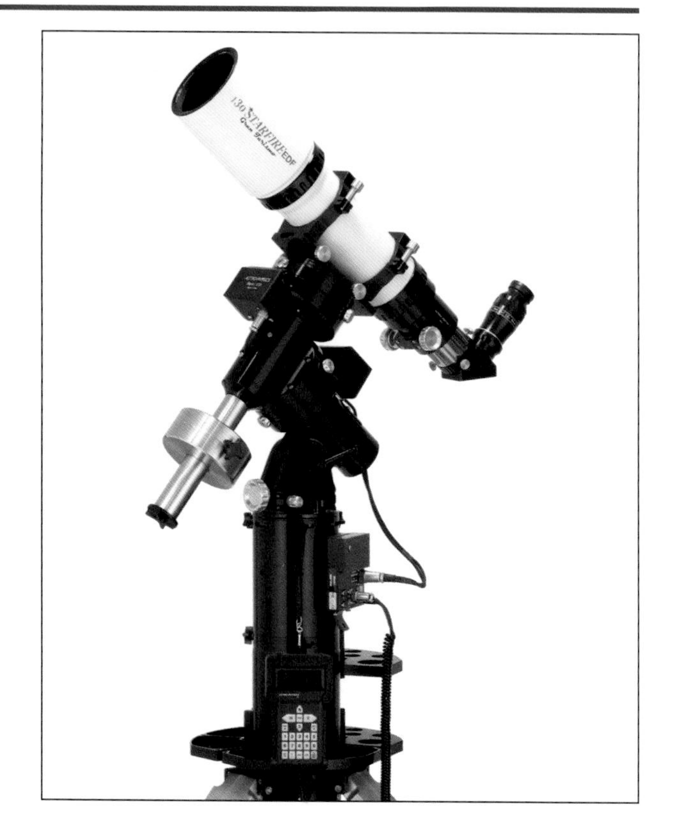

An *equatorial mount* also lets you swing the telescope along two axes. In this case, one axis moves the scope parallel to the celestial equator and the other moves in a perpendicular direction. In this way, they sweep along lines of right ascension and declination, respectively. An equatorial mount has to be lined up with the celestial pole, although a go-to system accomplishes this without a hassle.

TIP

Cool Down Your Telescope

No telescope will function perfectly immediately after you bring it from a warm house into the cool night air. If you give your scope time to cool to the ambient air temperature, you'll be rewarded with sharper views.

Eyepieces

Without a doubt, eyepieces are the most critical accessory for any telescope owner. Without an eyepiece, you won't see a thing. The telescope's main optics create an image of a celestial object, and the eyepiece magnifies that image. If you have an inferior eyepiece, the final image won't match your expectations. Most telescopes come with only one eyepiece, so you'll want to purchase others so that you have a range of magnifications available.

Eyepieces come in a wide range of sizes. One dimension has to do with the *barrel diameter*—the size of the hole in your telescope where the eyepiece goes. The most common barrel diameters are 1.25 and 2 inches. The other important dimension is the eyepiece's *focal length.* This determines how much an eyepiece will magnify the view on a given scope.

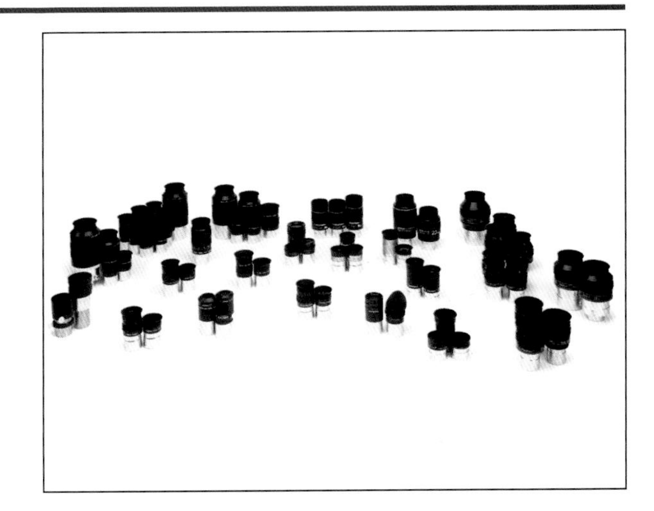

These three eyepieces have different focal lengths (you can tell by their heights). Focal length determines *magnification.* Magnification equals your telescope's focal length divided by the eyepiece's focal length. So, if your scope has a focal length of 2,000mm, a 40mm eyepiece delivers 2,000/40 = 50×. A 20mm eyepiece yields 100×, and a 10mm eyepiece gives 200×. Put those same eyepieces in a 1,000mm telescope, and the magnifications would be halved.

TIP

Beware of Empty Magnification

Occasionally, you'll hear a telescope touted as delivering 500× or so. Steer clear. Most such claims come from bargain-basement telescopes that contain a ridiculously short-focal-length eyepiece. Such eyepieces rarely give a view with any detail.

Filters

Filters rank just below eyepieces on any observer's list of accessories. Filters come in three varieties: *solar*, *planetary*, and *deep-sky*. Because you can't view the Sun directly without a filter, solar filters are discussed in detail in Chapter 8. Planetary filters simply enhance contrast and detail when viewing the planets (particularly Mars and Jupiter). Several kinds of deep-sky or nebula filters pass specific wavelengths emitted by nebulas while blocking all else.

If you want to pick out subtle detail on the surface of Mars, or see Jupiter's Great Red Spot more clearly, there's a filter for you. A light red filter (Wratten 23A), for example, will pump up the contrast between darker and lighter regions on Mars. And a light green filter (Wratten 56) brings out Jupiter's dark cloud belts and the Red Spot. Be warned that planetary filters make subtle improvements, and beginning observers may not notice a difference.

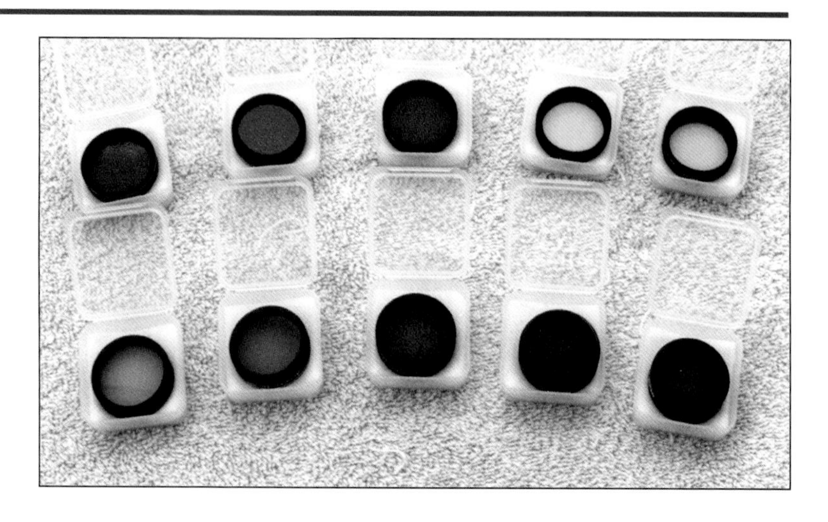

You don't have to be an experienced deep-sky observer to see the difference a nebula filter makes. By passing only the light emitted by emission nebulas, planetary nebulas, or supernova remnants, these filters darken the sky dramatically while letting the object of interest shine through. A good deep-sky filter can turn your suburban backyard into a dark-sky site—at least when it comes to viewing nebulas.

Say Goodbye to Dew

As the temperature drops during a long observing session, often your telescope will drop below the dew point and moisture will start collecting on optical surfaces. Bring along a dew cap, some heater coils, or just a hair dryer to keep the dew at bay.

Other Accessories

The number of accessories on the market today testifies to the ingenuity of the observing community. There's nowhere near enough room to highlight all the good products, but here are three.

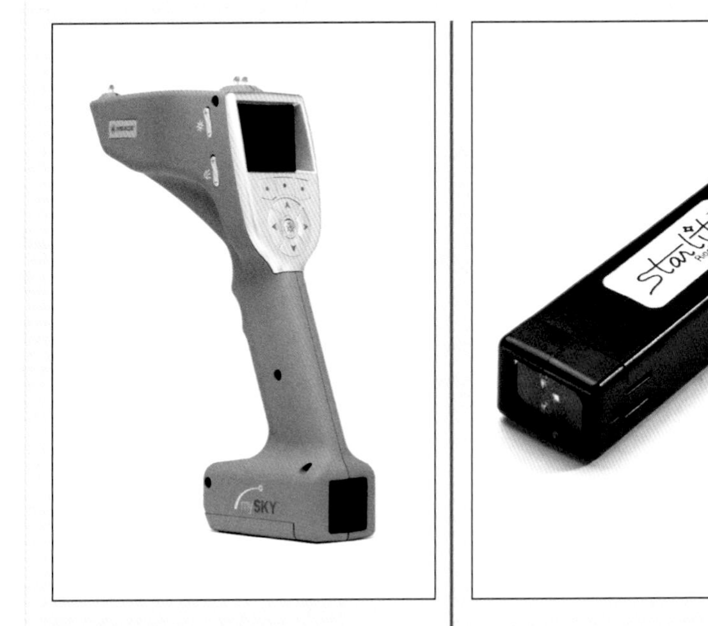

Curious what that object is you're looking at? Point either Meade's MySky or Celestron's SkyScout at the spot, and it will tell you.

If you spend much time observing at a dark site, you don't want to ruin your dark adaptation by using a white flashlight. Instead, buy a red flashlight or make one by taping red cellophane over your normal flashlight.

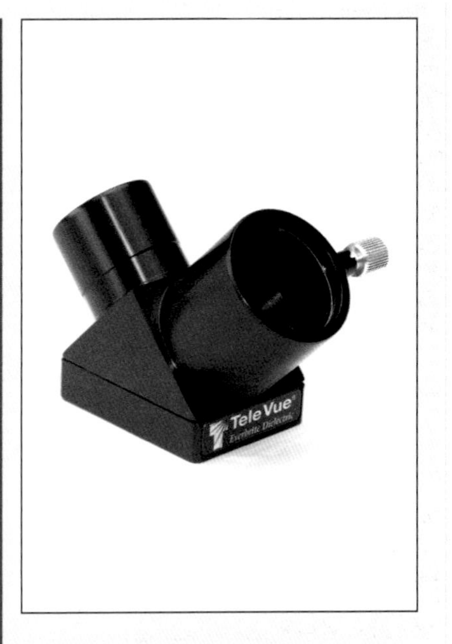

If you've ever had to lean over your telescope awkwardly to look into the eyepiece, you'll want a *star diagonal*. This device redirects the light from your telescope at a 90° angle so you can reach the eyepiece with ease.

chapter 3

View the Naked-Eye Sky

People getting started in astronomy often think they must purchase a telescope. Not true. You can begin to appreciate the sky with nothing more than your eyes. Some intriguing displays show up only in daylight, although you have to remind yourself to look up to see them. Other vistas are the province of night.

Some of these sights arise in Earth's atmosphere. Rainbows, auroras, and meteors, for example, occur within a few dozen miles of us. All of these trace their origins to celestial objects, however. Farther afield, solar system bodies and our galaxy's stars can keep naked-eye observers active and content for years to come.

Atmospheric Effects on Sunlight

After traveling 93 million miles unimpeded, sunlight has to traverse a dozen miles of air before reaching our eyes. That last little bit can play tricks and create beauty. Water droplets and ice crystals bend and reflect sunlight into rainbows, halos, and a variety of other colorful shapes. So remember: Skygazing doesn't have to be just a nighttime endeavor. It pays to keep looking up even during daytime.

High in the atmosphere, the six-sided ice crystals in cirrus clouds *refract* (or bend) light in the same way a prism does. The effect is to create a halo around the Sun with a radius of 22°. This is close enough that people typically don't notice a solar halo unless a building or some other object blocks the Sun. Under good conditions, you can see color in the halo, with red on the inside of the arc. A halo with the same radius can also form around a full Moon.

When the Sun hangs low in the sky and cirrus clouds abound, the 22° solar halo can brighten significantly at the Sun's elevation. This creates either a single or a matching pair of *sun dogs,* depending on where the clouds lie. (People occasionally call these features *mock suns* or *parhelia*.) Sun dogs occasionally show color, as in this photo, but they often appear as a diffuse white light.

Raindrops and sunlight combine to create one of the daytime sky's most impressive sights. If the Sun is shining during a storm, face away from the Sun and you should see at least one rainbow. Sunlight enters the near side of the droplet and gets refracted by the water, breaking the light into its component colors. This light then reflects off the back of the drop and exits through the front, where it once again gets refracted.

The light emerges from the raindrop at a 42° angle from where it entered. So, the Sun must be lower than 42° altitude for you to see a rainbow, and the bow has a 42° radius around a point exactly opposite the Sun. Because red light gets refracted less than violet, the outside of the rainbow appears red. If the sunlight is particularly strong, the light reflects a second time within the droplet and gives rise to a secondary rainbow. The secondary bow has a radius of 51° and glows fainter than the primary bow. (The sky between the bows appears significantly darker than the neighboring sky.) The colors in the secondary bow appear in reverse order.

Sunrise and Sunset

To many astronomy enthusiasts, few sights are as welcome as the Sun dipping below the horizon to herald the approaching night. But sunset, along with sunrise, offers its own visual pleasures. The sky's changing color as the Sun nears the horizon, and during twilight, can excite skygazers as much as the night sky itself.

Earth's midday sky appears blue because the tiny air molecules in our atmosphere most effectively scatter short-wavelength blue light. As the Sun nears the horizon, however, sunlight must pass through ever-increasing amounts of air. Eventually all the blue gets scattered out, and we're left with orange and red. These saturate the sky and any nearby clouds, giving us the characteristic colors of sunrise and sunset.

As the Sun nears the horizon, you'll often notice the bloated reddish orb appears squashed. That's another effect of Earth's atmosphere. Light from the bottom of the Sun has to pass through more air—and thus gets bent more—than light from the Sun's top. The net result: The top and bottom of the Sun appear closer together than they normally would, and the Sun's middle seems to expand.

Like a searchlight beamed from the Sun, a *Sun pillar* presents another compeiling demonstration of what atmospheric ice crystals can do. When the air is still and the Sun lies near the horizon, flat-sided ice crystals above or below the Sun reflect sunlight. More often than not, observers notice pillars above the Sun, and they can appear even when the Sun lies below the horizon. Because the crystals reflect sunlight, they typically appear the same color as the Sun.

As twilight deepens, sky colors often become more intense. Clouds at this time don't necessarily mean you'll have to call off nighttime observing. As the atmosphere cools after sunset, clouds often dissipate. If there's an event you want to see, don't let some sunset clouds discourage you from trying.

CONTINUED ON NEXT PAGE

FAQ

When does the Sun actually set?

By definition, the Sun sets when its upper limb disappears below the horizon. Because air bends sunlight, however, we still see the Sun after it should have dipped below the horizon. Sunset thus occurs a couple of minutes later than it would if Earth had no atmosphere.

Just because the Sun sets in the west, don't focus all your attention in that direction. If you happen to live where mountains climb in the east, those peaks will catch the Sun's rays long after you're in semi-darkness. Watch the interplay of light and shadow on the mountains, and stay alert for the constantly changing colors.

Looking east from the summit of Hawaii's Haleakala at sunset, the shadow of the great volcano makes a perfect triangle on the lower atmosphere. Many observers tend to get tunnel vision during an event's progress, whether the occasion is as mundane as a sunset or as rare as an eclipse. But it pays to keep looking around—sometimes the subtlest features are the ones you'll most remember.

As the last vestiges of twilight fade away, the night beckons. On a cloudless night like this, the sky really is the limit. Perhaps there'll be a close planetary conjunction to watch, or a bright aurora will play overhead. At the right time of year, it could be a meteor shower or the Milky Way in all its glory. On any clear night, you'll have plenty of sights to keep you occupied.

FAQ

How dark is twilight?

Astronomers recognize three stages of twilight. *Civil twilight* reigns while the Sun is less than 6° below the horizon. Terrestrial objects still appear distinct at this stage. *Nautical twilight* lasts until the Sun dips 12° below the horizon. In this period, the horizon appears indistinct. *Astronomical twilight*—and complete darkness—begins when the Sun is 18° below the horizon. The same stages occur, in reverse order, before sunrise.

Conjunctions

Almost every month, you can count on the Moon passing a bright planet. The Moon circles Earth once a month, but the planets cruise much more slowly. So, as long as Venus, Mars, Jupiter, and Saturn are visible, they will receive a monthly visit, called a *conjunction*. Each planet circles the sky at a different rate, and eventually they, too, approach one another. The best such conjunctions usually involve the two brightest planets, Venus and Jupiter.

The most spectacular planetary conjunctions almost always include Venus and Jupiter. Because these are the two brightest planets—indeed, they are the two brightest points of light in the sky—they grab people's attention whenever they pass near each other. In all such conjunctions, you can tell the two apart because Venus shines brighter.

If a two-planet conjunction is nice, three must be better. In this scene from June 27, 2005, Venus and Mercury lie so close to each other that they almost appear to merge. Venus is the brighter of the two, just to Mercury's right. As a bonus, Saturn lies 2° below the pair. (Gemini's two brightest stars, Castor and Pollux, sit to the right of the planetary trio.) Such close planetary conjunctions happen rarely enough that you should stay alert for them.

A crescent Moon adds to the night sky's drama whenever it passes a bright planet. Here, a waning crescent lies just below brilliant Venus, with Mercury perched a little higher. Because Venus and Mercury can never appear opposite the Sun in our sky, the Moon will always be a crescent when it passes these inner worlds. The outer planets have no such restriction on the passing Moon's phase. Unfortunately, the glare of a *gibbous* or Full Moon detracts from such scenes.

Constellations and Asterisms

Once the sky starts to darken, stars pop into view. Soon, you'll see patterns emerge. In springtime, the Big Dipper rides high. By summer and fall, the Summer Triangle stands out. But many of these patterns are not constellations. Instead, they are *asterisms*—recognizable groups of stars inside a constellation or combining elements from more than one constellation. Constellations themselves contain many more stars, often with patterns of their own.

The Summer Triangle asterism consists of three bright stars—Vega, Altair, and Deneb, in decreasing order of brightness—from three different constellations. Each is the luminary of its constellation, but bears no relationship to the other stars except in brightness and relative proximity. Those are hallmarks of an asterism, but not necessarily of a constellation.

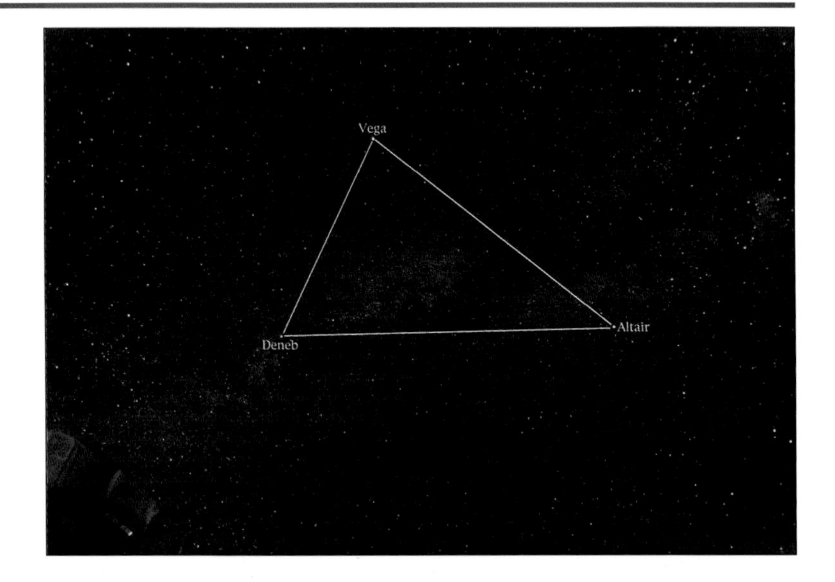

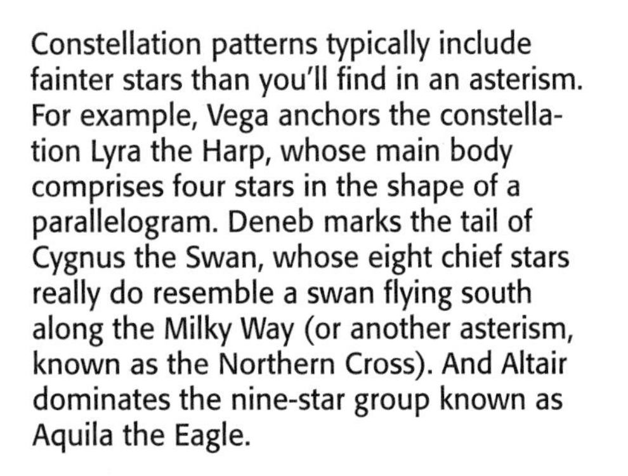
Constellation patterns typically include fainter stars than you'll find in an asterism. For example, Vega anchors the constellation Lyra the Harp, whose main body comprises four stars in the shape of a parallelogram. Deneb marks the tail of Cygnus the Swan, whose eight chief stars really do resemble a swan flying south along the Milky Way (or another asterism, known as the Northern Cross). And Altair dominates the nine-star group known as Aquila the Eagle.

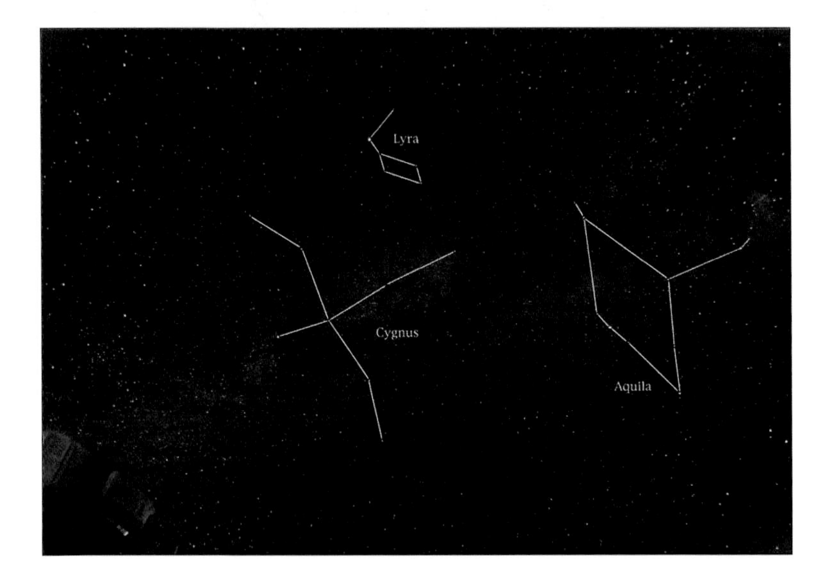

On a clear night from a dark site, you'll soon realize a lot more stars populate the heavens than those that form the constellation patterns. Are these really orphan stars without a constellation to call home? No. Astronomers have divided the entire sky into 88 constellations (see Appendix, "The 88 Constellations"). Every piece of real estate is accounted for. Even a faint star barely visible through a telescope, and perhaps with only an obscure catalog name, belongs to some constellation.

Auroral Lights

Few sights in the sky can compete with the aurora. Better known as the Northern or Southern Lights (depending on which hemisphere you live in), these eerie glows mesmerize people, particularly during an active display when the lights can ripple and whirl like curtains blowing in a stiff breeze. Auroras occur high in Earth's atmosphere, where atoms and molecules absorb energy from charged solar particles. And they occur most frequently at high latitudes.

The Sun ultimately causes the aurora. A steady stream of charged particles, called the solar wind, blows from the Sun. Our planet's magnetic field channels these particles toward Earth's magnetic poles. (Auroras occur at high latitudes because the north magnetic pole lies in arctic Canada and the south magnetic pole in Antarctica.) Particularly active displays follow coronal mass ejections, like the one seen here, which eject billions of tons of solar material at speeds of millions of miles per hour.

The most common aurora color is green. It comes from oxygen atoms that absorb some energy from the incoming solar particles. When the atoms return to a lower-energy state less than a second later, they emit the green color. Most of the green light comes from atoms at an altitude between 60 and 150 miles above Earth's surface.

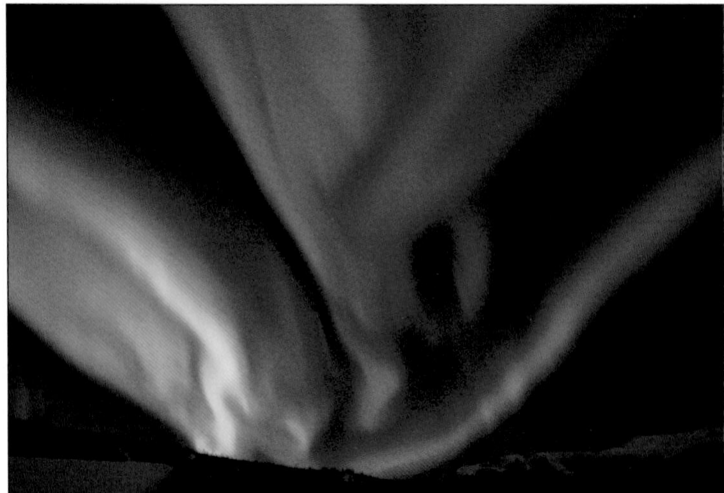

Although green may be the most common color, red shows up in some auroras. This color comes from two different elements: oxygen atoms that lie above 150 miles and nitrogen molecules below 60 miles. In this aurora, the red comes from oxygen and lies above and beyond the green emission. This image was taken in central Alaska, an aurora hot spot along with much of central and northern Canada and northern Scandinavia.

CONTINUED ON NEXT PAGE

When an intense auroral display passes overhead, it takes on a coronal form. In such a corona, the aurora's rays all seem to converge. This convergence point shows the local direction of Earth's magnetic field. After a corona forms, the aurora often quiets down, although the remaining patches of pulsating light certainly merit a look. Don't call it a night just yet—another active display may be on the way.

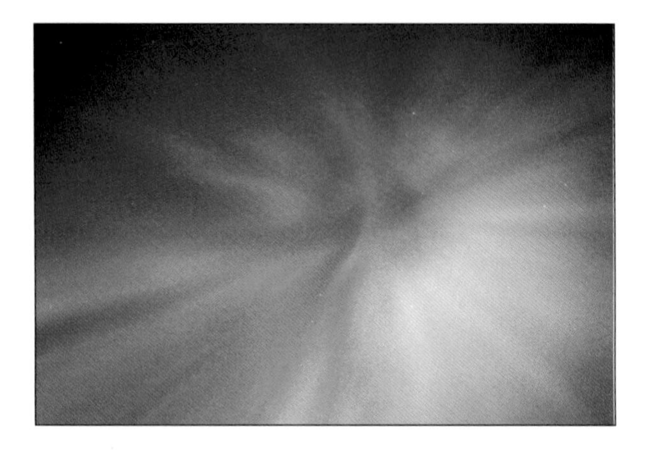

Rays of auroral light create a celestial curtain during this active aurora. The descriptive names of auroras include arcs, patches, bands, veils, and rays. Each of these forms can appear homogeneous, striated, or rayed, and can remain close to motionless or move wildly across the sky. An aurora's brightness can range from being nearly invisible to casting noticeable shadows.

This coronal display appears mostly green and red, but look closely and you'll see a hint of blue to purple. This color comes from ionized nitrogen molecules at altitudes below 60 miles. Auroral emissions occur in only three colors: green, red, and blue. Yet these primary colors can create virtually any color in the spectrum just by combining differing amounts of light from different altitudes. You never know what might turn up during an auroral display.

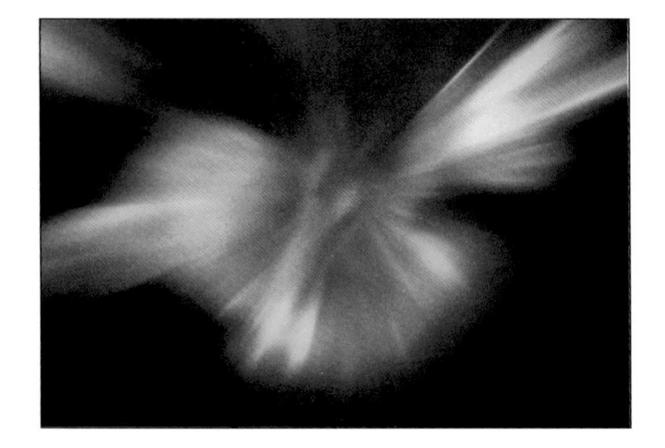

A bright aurora doesn't place the rest of the sky off limits. Even during this magnificent display, the stars of Cassiopeia and Perseus shine through easily. The question is: Why would you want to watch anything else? Active auroral displays occur so infrequently, at least at low latitudes, that you probably won't have much desire to view some distant deep-sky object.

TIP

Don't Miss the Next Auroral Display

Because the best auroral displays at mid-latitudes occur a couple of days after solar outbursts, scientists can predict when a good show might be in the offing. Good Web sites for predictions include www.spaceweather.com and www.spacew.com.

Meteors and Fireballs

Bits of debris run rampant through the inner solar system. When one of these pieces encounters Earth, friction with the atmosphere incinerates it. We see a flash of light—a *meteor.* At certain times of the year, Earth slams into a torrent of debris left behind by a passing comet. The resulting shower can produce a meteor or more every minute. Larger chunks of debris occasionally hit Earth, generating a brilliant flash known as a *fireball.*

When a comet comes close to the Sun, solar heat vaporizes some of its ices and bits of dust get released. After a comet passes the Sun repeatedly, particles spread out along the orbit and follow parallel tracks. If Earth's orbit intersects this stream, we'll see a meteor shower at the same time each year. All of a shower's meteors appear to radiate from a point. This is a perspective effect–the same reason train tracks or highway lanes appear to converge in the distance.

TIP

Seeing Meteors at Their Best

A few tips will have you seeing the most meteors. First, find an observing site far from city lights. Then dress warmly and lie back in a reclining lawn chair. Don't stare at the *radiant*–the point from which the meteors appear to emanate–meteors there have short trails and travel slowly. A good rule of thumb: look about 60° high and 60° from the radiant.

You don't need a major meteor shower to get a nice show. The meteor in this photo belongs to the minor Alpha Aurigid shower. As with most meteor showers, the Alpha Aurigids appear better in the after-midnight hours. As dawn approaches, Earth faces forward in its orbit around the Sun. Just as when you drive through a storm, and the front windshield picks up more raindrops than the rear, Earth sweeps up more meteors in the hours before dawn.

The Leonid meteor showers of the late 1990s and the beginning of the 21st century were the best of this past generation. In those years, Earth ran into exceptionally dense dust bands left behind by the periodic comet 55P/Tempel-Tuttle. Meteor rates jumped as high as 1,000 per hour—far more than the couple of dozen usually produced by the Leonids. In this image, more than a dozen meteors passed through the camera's field of view during a short exposure.

CONTINUED ON NEXT PAGE

Most Northern Hemisphere observers point to the August Perseids as the year's best meteor shower. The Perseids typically occur during warm weather and always deliver high rates. Under a clear, dark sky, observers often see an average of one meteor per minute. Keep in mind that this is an average, however. You might not see any meteors for five minutes, then spot five in the next minute. The Perseids appear to radiate from the constellation Perseus, hence their name.

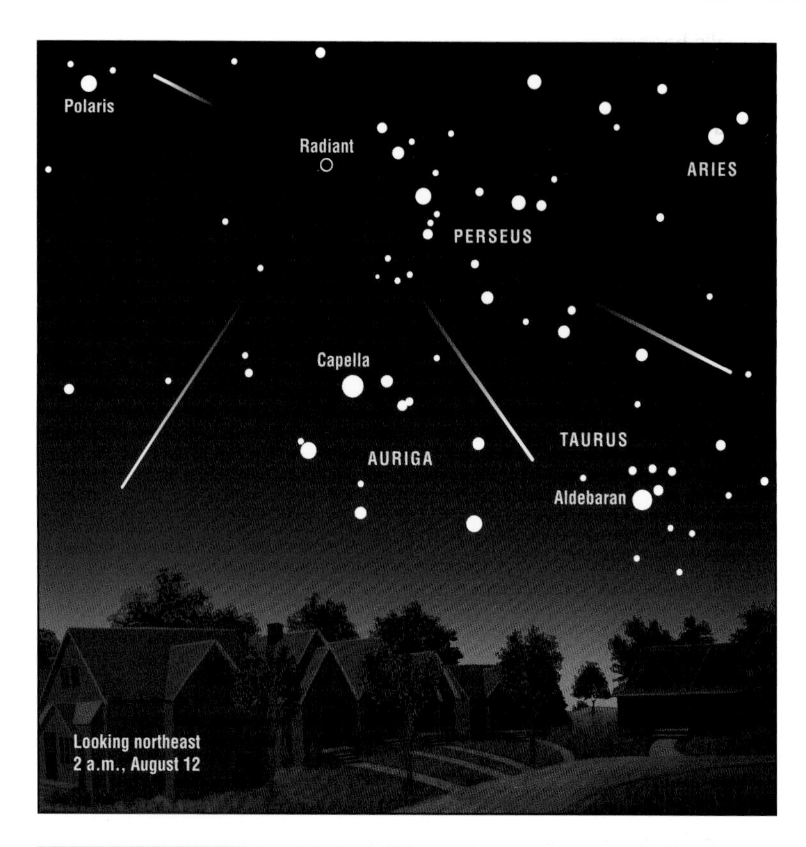

December's Geminid shower currently produces the highest average number of meteors. Unfortunately, it peaks at a cloudy and cold time of year, so fewer people observe the Geminid shower than they do the warm-weather Perseids. Geminid meteors derive not from a comet but from an asteroid, called Phaethon. This object likely was once a comet that lost all its ices after passing the Sun too many times. Geminid meteors appear to radiate from the constellation Gemini.

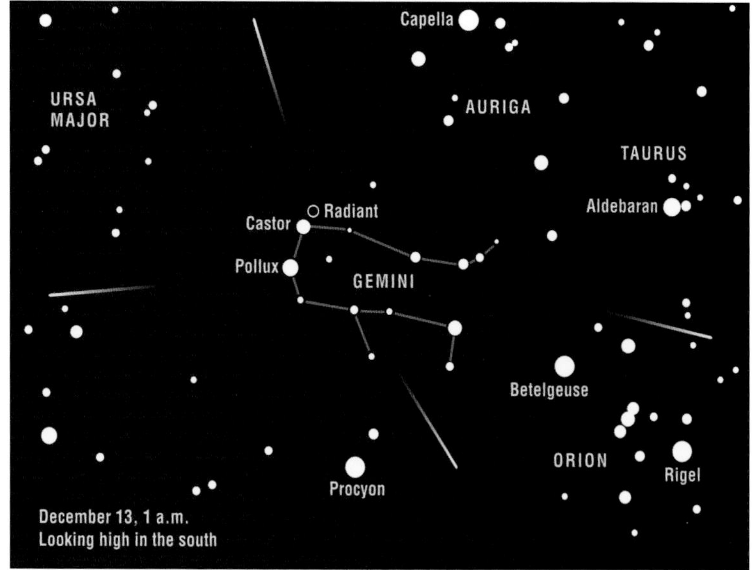

Fireballs happen without warning and don't belong to any shower. A typical shower meteor starts out as a particle no bigger than a grain of sand. A fireball enters the atmosphere as a pea-sized object or bigger. The added heft causes it to glow at least as bright as the planet Venus, the standard definition of a fireball. In this photo, a fireball appears in the sky above Alaska during a bright auroral display.

Major Meteor Showers		
Name	***Date of Peak***	***Maximum Rate (per hour)***
Quadrantids	January 4	120
Lyrids	April 22	20
Eta Aquarids	May 5	70
Southern Delta Aquarids	July 27	20
Perseids	August 12	100
Orionids	October 21	30
Leonids	November 17	20
Geminids	December 13	120

Note: *Dates can vary a day either way because of a leap year's effect. Maximum rate is for an observer under perfect conditions, which rarely happen.*

The zodiac encompasses a band centered on the *ecliptic* (the Sun's apparent path) and extending some 8° on both sides. That's wide enough that the Moon and the planets always lie within the zodiac. But there's more to the ecliptic and the zodiac than planets. A planet's visibility often depends on the angle the ecliptic makes to the horizon after sunset and before sunrise. And the ethereal zodiacal light appears only under favorable conditions.

The ease with which the inner planets appear depends largely on the angle between the ecliptic and the horizon. If the angle is steep, a planet's distance from the Sun translates into altitude. If the angle is shallow, the planet remains near the horizon. For observers at mid-latitudes, the steepest angles occur on spring evenings and autumn mornings. In the tropics, the angle is always steep. Here we see Venus shining brightly above the Moon from a Mexican beach.

A good way to tell the angle between the ecliptic and horizon is to look for a crescent Moon. If the Moon's cusps stick almost straight up, as they do here, the angle must be steep. (The cusps point away from the Sun, and so the Sun then must lie directly below the Moon.) At times like these, the crescent Moon stands as far above the horizon as it can. If the cusps parallel the horizon instead, you'd better wait before tackling the zodiacal light.

The zodiacal light appears as a pyramid of softly glowing light in the west after dusk or the east before dawn. It's slightly fainter than the brightest parts of the Milky Way, so you'll need a clear, moonless night to see it. And because it follows the zodiac, the light shows up best on spring evenings and autumn mornings, or from the tropics. Don't be fooled: Many people mistake the zodiacal light for the last vestiges of evening twilight or the first hint of dawn.

The Milky Way

The most spectacular naked-eye sight available on most clear nights has to be the Milky Way. Unfortunately, city dwellers and suburbanites alike miss out on this band of light that marks our home galaxy's disk. You need a dark sky to see the Milky Way in its glory. The best time to view it is summer, when the central regions in Sagittarius and Scorpius lie due south and the dusty clouds of Cygnus stand overhead. But anytime except spring will give a good view.

Astronomers think our galaxy would look like this if we could see it from above. Near the center lies a long stellar bar. From our perspective roughly halfway between the galaxy's center and edge, the bar remains hidden in visible light. What we see instead are the spiral arms. Most of the billions of stars in the arms lie so far away that they merge into a gentle glow—the Milky Way. This band appears brightest toward the galaxy's center.

Toward the center of the Milky Way in Sagittarius and Scorpius, billions of stars crowd together in the dense spiral arms. The thick dust lane visible here prevents the Milky Way from being even brighter, a godsend for observers interested in fainter deep-sky objects. Still, from parts of the Southern Hemisphere where the galaxy's central regions pass overhead, the Milky Way casts discernible shadows.

The Milky Way spans the sky from horizon to horizon in this early summer view from Hawaii. The galaxy's rich central regions lie toward the upper right in this photo, while the fainter, northern sections stretch to the left. Viewing the Milky Way from a dark site lets you envision our galaxy's three-dimensional shape. It's a sight worth traveling to a dark location to see.

chapter 4

Explore the Winter Sky

The nights may be cold, but the stars seem to sparkle with special intensity at this time of year. No other season boasts as many bright stars as winter does. Fully five of the sky's ten brightest stars shine prominently on winter evenings. And the brightest star of all—Sirius—twinkles madly low in the south.

The cold weather does make observing a little tricky, however. You'll need to dress warmly, and try to mix in some breaks to warm up between viewing sessions. On those nights when the sky clears, you'll be rewarded with vistas that are sure to send a chill down your spine.

Learn the Winter Constellations

Head outside on a dark night in the country, and the number of stars visible can seem overwhelming. A trip outdoors in the heart of a big city can prove equally disconcerting, with only a handful of stars shining through the haze. Neither situation will help much in your quest to learn the constellations (although, truth be told, you stand a better shot under a dark country sky). The all-sky map on the opposite page, and on the corresponding pages of the subsequent seasonal star maps, represents the stars as you would see them on a clear night from the suburbs. Such sky conditions will show you all the major constellations and many of the lesser ones.

On a winter's night, one constellation stands out particularly well: Orion the Hunter. It lies due south around midnight in mid-December. You can find it occupying the same position at 8 p.m. in mid-February. Orion boasts seven bright stars, but the most conspicuous are three closely spaced gems that all shine at 2nd magnitude. These form the Hunter's unmistakable belt. Two of Orion's stars shine even brighter. Ruddy Betelgeuse lies to the upper left of the belt, while blue-white Rigel lurks to the belt's lower right. A short line of stars descending from Orion's belt points directly to the magnificent Orion Nebula (M42).This stellar nursery—a place where stars are born from gas and dust—is one of the brightest visible from Earth.

Orion serves as a guide to many of winter's other wonders. If you extend the imaginary line that joins the three belt stars to the lower left, your eyes will land on brilliant Sirius in Canis Major. At magnitude –1.5, Sirius is the brightest star in the night sky. For observers at mid-northern latitudes, Sirius often twinkles crazily because it lies low in the sky and its light has to pass through thick layers of Earth's atmosphere.

If you extend the line joining the stars of Orion's belt to the upper right, your gaze will fall on Taurus the Bull and its bright star, Aldebaran. Attached to Aldebaran is a star cluster known as the Hyades, which forms the Bull's V-shaped face. An even more impressive star cluster, the Pleiades, sits above the Hyades. Be sure to view the Pleiades with your binoculars—you won't be disappointed.

Attached to Taurus, you'll find Auriga the Charioteer. This constellation's brightest star, Capella, passes nearly overhead on winter evenings. Following in its footsteps are a pair of slightly dimmer stars, Castor and Pollux. You should see a line of fainter stars extending from both these luminaries back toward Orion. These form the twin bodies of the constellation Gemini. Slightly below Castor and Pollux you'll see the brighter star Procyon. It's the main feature of the small and otherwise inconspicuous constellation Canis Minor.

Once you've mastered these main stars and constellations, the fainter winter groupings will be much easier to find. But before you peek ahead to our constellation close-ups, take one last, broad view of the winter sky. Start with Rigel, and then make a clockwise loop out of the bright stars. Pick up Sirius, then Procyon, then the pair of Pollux and Castor, followed by Capella, Aldebaran, and back to Rigel. You've just found one of the largest patterns of stars in the sky: "the Winter Hexagon."

This map shows the entire sky visible from 35° north latitude at approximately 1 a.m. on December 1, midnight on December 15, 11 p.m. on January 1, 10 p.m. on January 15, 9 p.m. on February 1, and 8 p.m. on February 15.

Discover the Jewels in Taurus

Taurus the Bull features one 1st-magnitude star, Aldebaran, and two standout star clusters best viewed with binoculars: the Hyades and the Pleiades. For those with a telescope, Taurus offers the Crab Nebula, one of the most spectacular supernova remnants in the sky.

The main body of Taurus the Bull is easy to trace. First, find the sprawling Hyades star cluster. The brightest stars in this group take the shape of the letter "V." The 4th-magnitude star Gamma (γ) Tauri marks the V's tip. Aldebaran represents the fiery eye of the Bull. Although this star appears to belong to the Hyades, it lies in the foreground, at less than half the cluster's distance.

If you extend the arms of the Hyades's V eastward, you'll come to Beta (β) and Zeta (ζ) Tauri. These two stars mark the tips of the Bull's horns. The Crab Nebula (M1) lies just 1° northwest of Zeta. But the finest deep-sky object in Taurus has to be the Pleiades star cluster (M45). It stands at the Bull's shoulder.

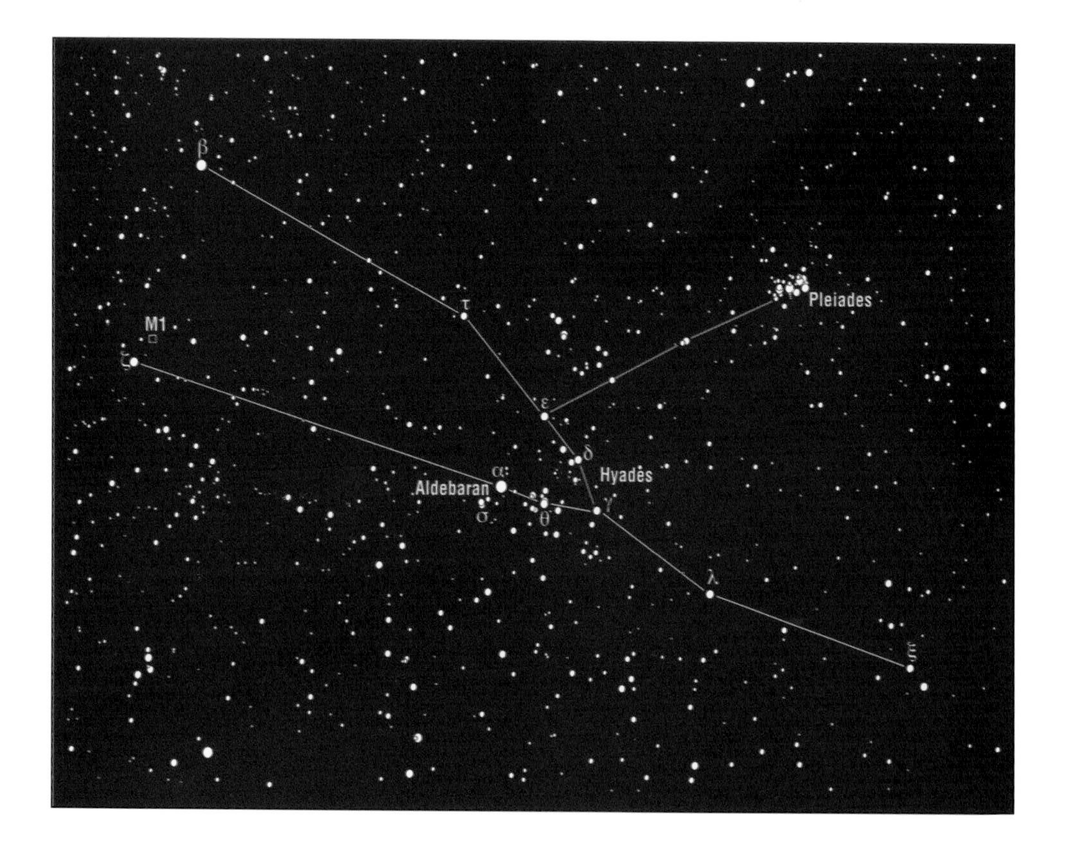

Taurus stands highest in the southern sky around 11 p.m. in mid-December and 7 p.m. in mid-February. From mid-northern latitudes, this constellation rises and sets about seven hours before and after these times. When Taurus lies due south, it appears as it does in the star chart and photo on these pages, with the V of the Hyades cluster pointing to the right (west). As the Bull sinks toward the western horizon later at night, the tip of the V leads the way.

To the naked eye, colorful stars are few and far between. That's because most stars glow too dimly to excite our eyes' color receptors. In Taurus, only 1st-magnitude Aldebaran shines bright enough to show a slight orange color. Because binoculars and telescopes gather more light, they show star colors more vividly. In Taurus, you'll see a smattering of blue, yellow, and orange stars.

CONTINUED ON NEXT PAGE

The V-shaped Hyades star cluster spans more than 5°—making it so big that it looks like a cluster only with the naked eye or binoculars. The cluster spreads out so much because it lies relatively close to Earth, "only" 150 light-years away. Aldebaran, which lies along the same line of sight as the Hyades, appears much brighter in part because it lies just 65 light-years from us.

Although the Hyades doesn't look like a cluster through a telescope, a scope still reveals some worthy details. In particular, the Hyades has several pleasing double stars. Among the best for small telescopes are Theta (θ) and Sigma (σ) Tauri.

The Pleiades star cluster (M45) provides a great target for the naked eye, binoculars, or a small telescope at low power. It lies about 440 light-years from Earth and appears more compact than the Hyades. The Pleiades also serves as a good test of eyesight. To some people, the cluster appears as a fuzzy patch of light to the naked eye. Most observers see six stars in the shape of a tiny dipper. And a few lucky souls claim to see a dozen or more stars on the best nights.

On July 4, 1054, our ancestors looked to the sky and saw a brilliant point of light that rivaled the brightness of a crescent Moon. It was a supernova—the death throes of a massive star that blew itself apart. Now, more than 950 years later, we can see the tattered remnants of that star. Because sketches made by 19th-century observers bore some resemblance to a crab, the object was christened the Crab Nebula (M1). You can locate the nebula about 1° northwest of the 3rd-magnitude star Zeta (ζ) Tauri. Through a small telescope, the Crab may look disappointing: an amorphous blob without much structure. A large telescope at a dark site begins to show the Crab's filaments.

To the sharp eyes of the Hubble Space Telescope, the Crab Nebula explodes with detail. In the time since the star exploded in the constellation Taurus, it has expanded at a speed of about 3 million miles per hour, and now spans approximately 12 light-years. The remnant itself lies some 6,500 light-years from Earth.

In this image, the orange filaments represent the outer layers of the star that exploded. They consist largely of hydrogen gas. The core of the exploded star lies near the heart of the nebula. It is a neutron star—a star with about twice the mass of the Sun crammed into a region no bigger than a city. A teaspoon of neutron-star material would weigh millions of tons. The Crab Nebula's neutron star rotates about 30 times every second, and supplies the energy that keeps the surrounding gas glowing.

Survey the Wonders in Orion

Orion the Hunter boasts two of the sky's ten brightest stars, Betelgeuse and Rigel, which sandwich the three slightly fainter stars of Orion's belt. The constellation's dazzling stars tell only part of its story, however. Just as impressive is the Orion Nebula—a star-forming region with few rivals. The Hunter boasts several other glowing gas clouds, including M78 and NGC 2024.

Most people recognize Orion the Hunter as soon as it's pointed out. With seven stars glowing at 2nd magnitude or brighter, it pierces the pall of light pollution even from a city. To trace the Hunter's shape, start with the three equally bright stars that mark the Hunter's belt. These closely spaced stars define the figure's central part. Scan north of the belt to find Orion's two shoulder stars: Betelgeuse and Gamma (γ) Orionis. Then head south of the belt to pick up the two knees: Rigel and Kappa (κ) Orionis. The Orion Nebula (M42) lies between the belt and the knees, at the end of a string of stars that forms the Hunter's sword. The reflection nebula M78 lies a little northeast of the belt.

The rest of the Hunter requires a darker sky to see. His shield consists of six stars to the body's west, while his upraised club extends north from Betelgeuse. In mythology, the Hunter is battling Taurus the Bull, a conflict also played out in the sky. Watch on any winter's night as Orion chases Taurus across the sky.

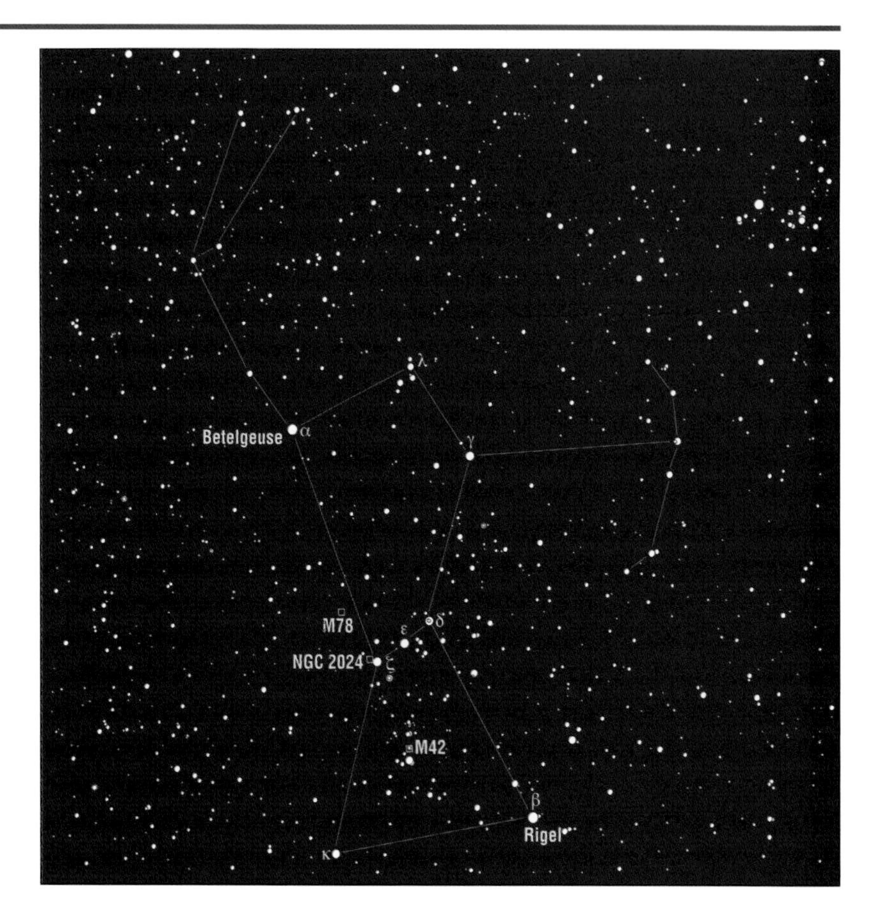

Orion climbs to its highest position in the south around midnight in mid-December and 8 p.m. in mid-February. The constellation rises and sets approximately six hours before and after these times. Orion takes on its classic form, with the shoulders placed directly above the knees, when it lies due south. Its shape looks decidedly different when the constellation lies closer to the eastern or western horizon, but the three belt stars will help you pick it out anyway.

Most of the stars in Orion sport a distinct blue-white hue. That color appears most noticeable with brilliant Rigel. But the most conspicuous color has to be Betelgeuse's orange-red glow. The other reddish glows in this photo represent star-forming regions. You'll need some optical aid to see even a hint of color in the Orion Nebula, the bright, pinkish region below the belt.

CONTINUED ON NEXT PAGE

Survey the Wonders in Orion *(continued)*

The three blue-white stars of Orion's belt stand out to the naked eye, but the region surrounding it holds interest for those with telescopes. The Flame Nebula (NGC 2024) sits just east (to the left) of the belt's easternmost star, Zeta (ζ) Orionis. Although the Flame Nebula shines brightly with the glow from many young stars, it's hard to pick out from the glare of 2nd-magnitude Zeta. Your best bet: Increase your telescope's magnification to 100× or so, and nudge it so that Zeta lies outside your field of view.

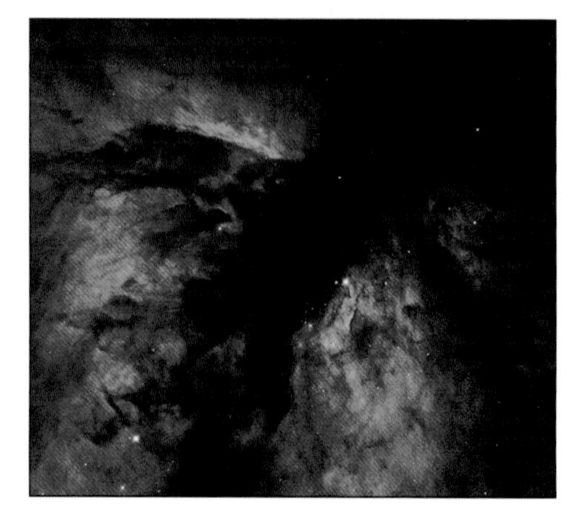

Just 2.5° northeast of Zeta Orionis lies one of the sky's brightest reflection nebulas, M78. A reflection nebula doesn't emit light on its own. Rather, it consists mostly of dust particles, which scatter and reflect light from nearby stars. In M78's case, the culprits are hot, blue stars, and so the nebula takes on a characteristic bluish tinge in long-exposure photographs. Although it's easy to locate with a small telescope, M78 shows little detail.

The Orion Nebula (M42) glows bright enough to see with the naked eye—it's the fuzzy, middle "star" in the Hunter's sword. Binoculars reveal this star-forming region's gaseous nature, while a telescope shows incredible detail across its face, which covers an area much bigger than the Full Moon. Even through a 2.4-inch telescope, you'll see a group of four bright stars located at the nebula's heart. This is the Trapezium, a quartet of the youngest, hottest stars in our galaxy. Thousands more stars of all sizes have started to condense from this cloud.

Of all the stellar nurseries visible easily from the Northern Hemisphere, none matches the Orion Nebula. This huge swirl of gas and dust looks spectacular in part because it lies relatively near Earth, at a distance of 1,500 light-years. It already has given birth to thousands of stars, and the nebula contains enough material to create thousands more. The cloud appears red because high-energy radiation from the hot, young stars inside it excites hydrogen gas to glow that color.

This Hubble Space Telescope image also reveals a smaller star factory at top left. Although it belongs to the same complex as M42, it has its own designation: M43. Small dust disks surround many of the newly forming stars. Planets may one day coalesce from these disks, creating new solar systems.

View the Gems in Gemini

Gemini the Twins' two brightest stars, Castor and Pollux, anchor the constellation's eastern end. From there, scan to Gemini's opposite corner to locate its premier star cluster, M35. Although you can spot it with the naked eye under a dark sky, it's a treat through binoculars or a telescope. You'll need a telescope to view the Eskimo Nebula, a gas cloud ejected by a Sun-like star near the end of its life.

It takes a little imagination to see twins in the stars of Gemini. First, start with Castor and Pollux near the constellation's eastern edge. First-magnitude Pollux shines a little brighter than Castor. From each of these luminaries, a string of fainter stars trails to the southwest. The Twins stand on feet planted firmly in the Milky Way. If you have trouble seeing twin brothers in these stars, think of Gemini as a long box.

The best star clusters in Gemini naturally lie in the Milky Way, at the constellation's feet. You can find M35 and its companion, NGC 2158, just above the northern foot, represented by the stars Mu (μ) and Eta (η) Geminorum. The Eskimo Nebula (NGC 2392) stands on the opposite side of Gemini. Look for it southeast of the 4th-magnitude star Delta (δ) Geminorum.

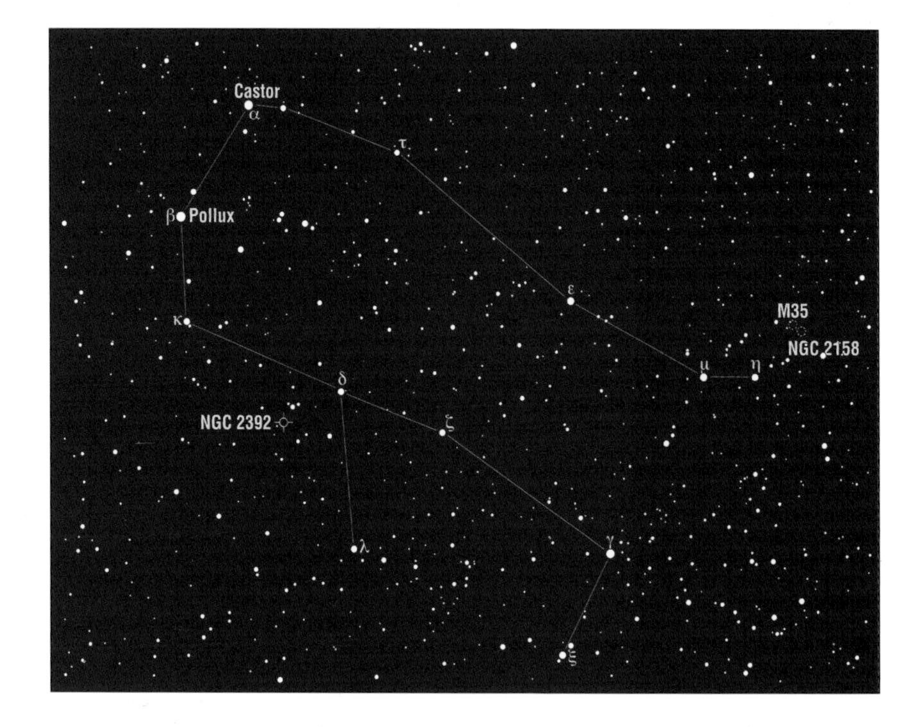

As the stars move from east to west during the course of the night, it takes longer for Gemini to reach its prime position in the south than it does for either Taurus or Orion. Gemini doesn't peak until nearly 2 a.m. in mid-December and 10 p.m. in mid-February. (From the southern United States, Gemini passes nearly overhead at these times.) It takes about eight hours from the time the constellation rises for it to reach its highest point, and another eight hours pass before it sets.

As the easternmost of the major winter constellations, Gemini remains visible in the west on evenings well into spring. As it dips toward the horizon, the Twins descend feet first. Castor stares down on Earth with an icy winter whiteness touched with a hint of blue. Pollux, on the other hand, exudes a warmer, yellow-orange hue, perhaps a harbinger of spring's approach. None of Gemini's other stars shine bright enough to show color to the naked eye.

CONTINUED ON NEXT PAGE

View the Gems in Gemini *(continued)*

With a name like Gemini the Twins, this constellation might already have you seeing double. And Gemini's second-brightest star, Castor, only reinforces this impression. If you turn a small telescope on Castor, you'll see that it splits into a pair of stars. Castor was the first star astronomers found that made a physical pair, where two stars orbit each other. It takes nearly 500 years for the stars to complete one revolution. The distance between the two is now growing, and so they are becoming easier to split. A third, much fainter star also belongs to this group. And perhaps most amazing, all three stars have close companions of their own, making Castor a system with six stars.

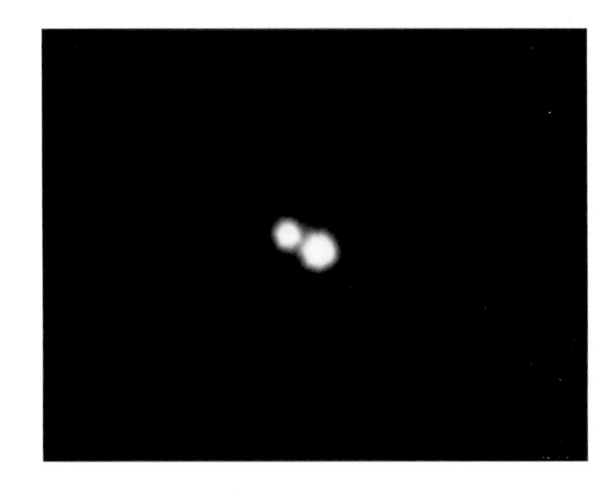

Another Gemini pair shows up near its western border with Taurus. Under a dark sky, you can spot the star cluster M35 (see left photo) without optical aid. Through binoculars or a telescope, it becomes spectacular. A telescope reveals a few dozen bright stars near the cluster's center. In all, the cluster holds some 200 stars. At low power, you should see a fuzzy ball southwest of M35 (lower right of right photo). This is NGC 2158, which appears smaller than M35 because it lies some four times farther away.

With a name like the Eskimo Nebula (NGC 2392), this planetary nebula should feel right at home in the winter sky. The object got its nickname because, through small telescopes, it vaguely resembles a face surrounded by a fur parka. The Eskimo shows a slight bluish cast in many scopes, and its disk divides into two concentric shells. A dark ring separates the brighter, inner shell from the outer shell. You also should see a conspicuous star at the center of the gaseous glow.

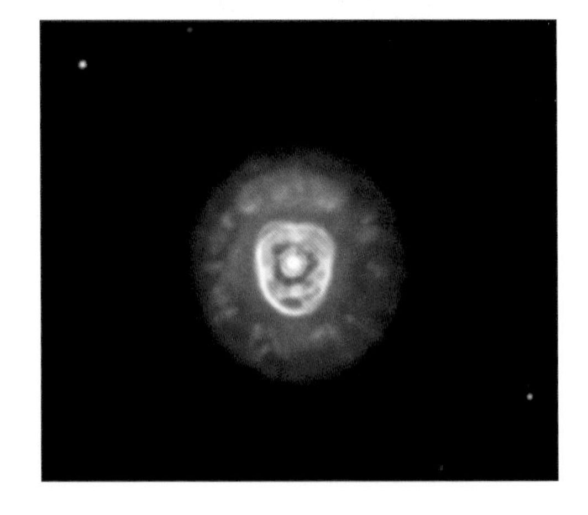

The Eskimo Nebula showcases a brief but dramatic stage near the end of a Sun-like star's life. Under the watchful eye of the Hubble Space Telescope, the Eskimo shows tremendous detail. Several tightly bunched gaseous shells occupy the nebula's inner part, while the outer part resolves into myriad streamers. These streamers—the parka's "fur"—resemble giant comets pointing toward the nebula's center.

A star with a mass near that of the Sun will eventually run out of hydrogen fuel in its core. (No need to worry, it won't happen to the Sun for another 5 billion years.) When it does, its outer layers swell and the star becomes a red giant. Repeated pulsations eventually puff off the giant star's outer layers, giving rise to the intricate shells seen in the Eskimo Nebula. The star's core, now stripped bare, glows white hot at the nebula's center. This white dwarf emits a lot of high-energy radiation, which energizes the surrounding gas and causes it to glow.

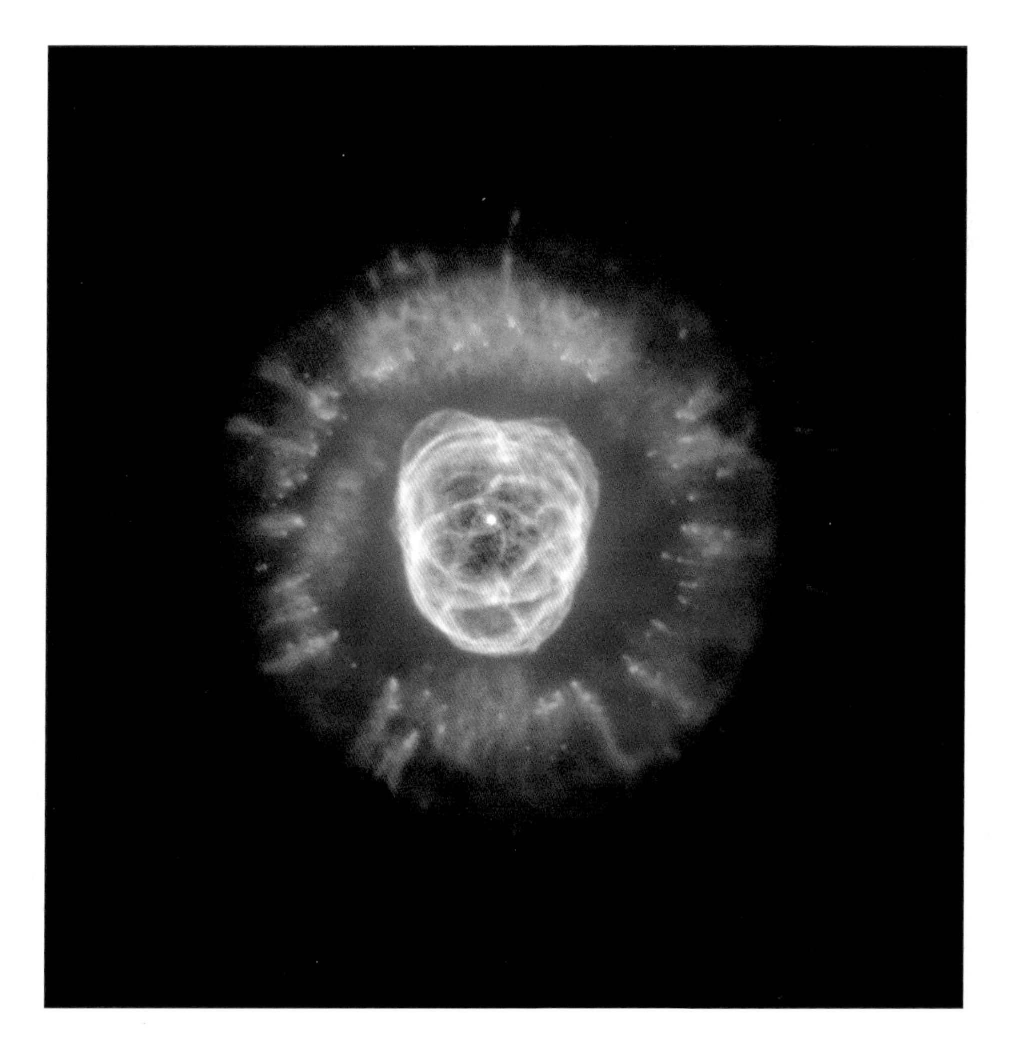

chapter 5

Explore the Spring Sky

Warm spring breezes seem to drive the brilliant stars of winter away. As Gemini, the last of the prominent winter constellations, dips below the western horizon, the night sky takes on a low-key appearance. Yet plenty of treasures still await you.

The Big Dipper—the most conspicuous star pattern in the northern sky—dominates spring evenings. And the greatest collection of galaxies visible through a small telescope appears at its best this time of year. So take advantage of the warmer weather, and track down springtime's distant treats.

Learn the Spring Constellations

Finding your way around the night sky is a little like getting your bearings in an unfamiliar city. The best-laid plans require a good map and at least one conveniently placed, easy-to-read sign. The all-sky map at right takes care of the first necessity. And the best sign in the heavens—the Big Dipper—passes nearly overhead on spring evenings.

The Big Dipper's seven stars–four make the bowl and three make the handle–create the most familiar asterism in the sky. An *asterism* is a group of stars that forms an identifiable shape but does not, by itself, represent a constellation. In the Big Dipper's case, the seven stars make up only a small part of the constellation Ursa Major, the Great Bear. The Big Dipper helps people navigate the sky not only because it's recognizable, but also because it remains visible most of the night nearly all year long. In fact, from the northern United States, Canada, and most of Europe, this group of stars never sets.

Let's begin our tour with the two stars that form the end of the Big Dipper's bowl. Extend the line joining these "pointer stars" to the north and your eyes will land on Polaris, the North Star. This 2nd-magnitude star stands out not because it's so bright, but because it never seems to move. Earth's axis of rotation points toward Polaris. As a result, all the sky's stars rotate around this star. Polaris belongs to the Little Dipper in Ursa Minor, where it marks the end of the Dipper's handle. The rest of the Little Dipper is hard to trace, with only the bowl's two end stars easy to spot. On spring evenings, the Little Dipper arcs upward from Polaris. A long, meandering line of stars lies between the two Dippers. These stars represent the body of Draco the Dragon.

Head back to the Big Dipper, and extend the line joining the pointer stars in the opposite direction. Your eyes will fall on Leo the Lion. Although Leo's stars form a conspicuous pattern, they won't bring a lion to mind. Look for a backward question mark anchored by the 1st-magnitude star Regulus. To the east of the question mark lie three stars that form a right triangle; these represent the Lion's hindquarters. Halfway between Leo and the setting stars of Gemini is Cancer the Crab. This inconspicuous grouping's main claim to fame is the presence of the bright star cluster known as the Beehive (M44). You can see the Beehive with your naked eye under a dark sky. Binoculars or a small telescope will show you dozens of stars packed together.

We aren't through with the Big Dipper yet. If you follow the curve of the Dipper's handle away from the bowl, you'll find spring's brightest star, Arcturus. This orange-colored gem dominates the kite-shaped constellation known as Boötes the Herdsman. Extend the arc of the Dipper's handle farther and you'll come to 1st-magnitude Spica, the brightest star in Virgo the Maiden. This constellation houses the center of the Virgo Cluster, the largest set of galaxies visible through a small telescope. An old saying will help you remember the bright stars the Dipper's handle points to: "Arc to Arcturus, then drive a spike to Spica."

South of Cancer, Leo, and Virgo, swims Hydra the Water Snake, the longest constellation of all. If not for its distinctive oblong head (south of Cancer) and the 2nd-magnitude star Alphard, Hydra wouldn't stand out much at all. The same can't be said for a conspicuous group of four stars that rides on the Water Snake's back: Corvus the Crow. Corvus shows up nicely even from the suburbs.

Our final spring constellations lie northeast of Arcturus. A compact semicircle of stars forms Corona Borealis, the Northern Crown. And next to it resides the sprawling Hercules the Hero. Hercules boasts the northern sky's brightest globular star cluster, M13. You can glimpse M13 with your naked eye under a dark sky, but you'll need to turn a telescope on it to appreciate it fully.

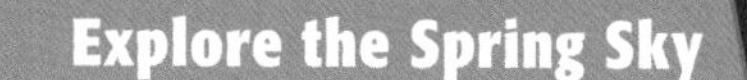

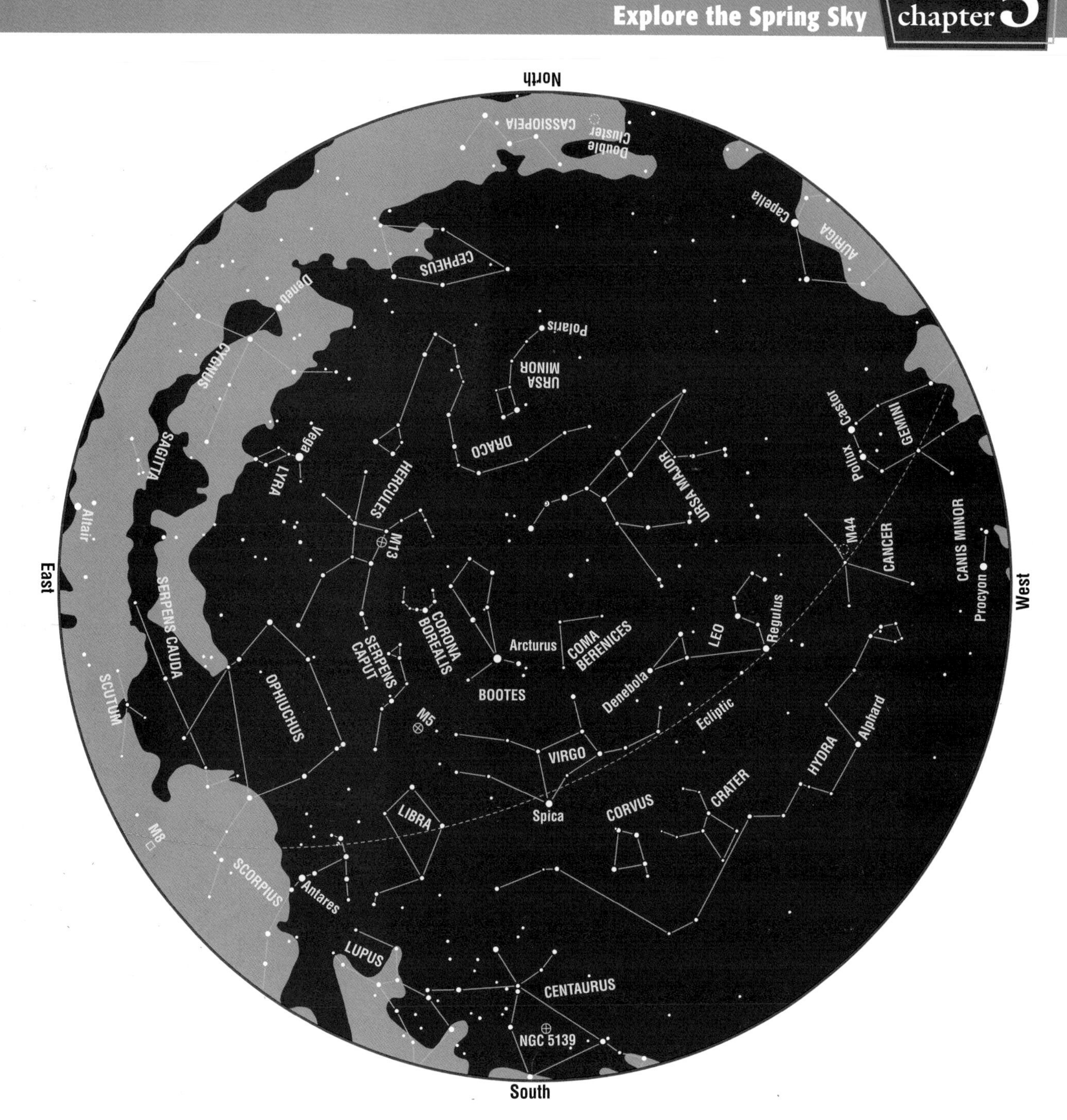

This map shows the entire sky visible from 35° north latitude at approximately 3 a.m. on March 1, 3 a.m. on March 15, 2 a.m. on April 1, 1 a.m. on April 15, midnight on May 1, and 11 p.m. on May 15.

View the Marvels in Leo

Leo the Lion features one 1st-magnitude star, Regulus, and one 2nd-magnitude star, Denebola. But Leo derives most of its fame from its plentiful galaxies. Five bright galaxies from Charles Messier's catalog—M65, M66, M95, M96, and M105—call Leo home.

Leo the Lion divides into two distinctive sections. The Lion's head and front looks like a backward question mark, although some see a gardener's sickle instead. The bright star Regulus forms the base of the question mark and represents the Lion's heart. Well east of this group, you'll see the Lion's reclining hindquarters. The three brightest stars here take the shape of a right triangle, with Denebola as the tail.

Although galaxies rule the Lion's lair, don't miss the pretty double star Algieba, which lies in the midst of the question mark. Leo's brightest galaxies all lie south of the Lion's body. M95, M96, and M105 cluster together below the beast's midsection, while M65, M66, and their worthy companion, NGC 3628, sit just south of the Lion's hind legs.

Leo reaches its highest point in the south around midnight local daylight time in mid-March and 8 p.m. in mid-May. From mid-northern latitudes, this constellation rises and sets approximately seven hours before and after these times. The Lion, even though artists typically depict him reclining, moves as a stalking cat should, with his head leading the way. When Leo stands highest in the south, his haunches parallel the horizon.

To most people, Regulus is the only star in Leo that shows significant color; it sports a noticeable blue-white hue. Other star colors come out with the increased light-gathering power of a telescope. If you look at the bright double star Algieba, for example, both components appear distinctly yellow. The relatively small size and faintness of Leo's myriad galaxies keep them from showing up in this photograph.

CONTINUED ON NEXT PAGE

View the Marvels in Leo *(continued)*

Spiral galaxy M95 resides in south-central Leo, next to two companion galaxies: M96 and M105. M95 glows at 10th magnitude, and so you'll need a telescope of modest aperture (8 inches or larger) to see much detail. Look for a bright center surrounded by an elongated glow. The spiral arms appear as a slightly brighter ring at the halo's edge. Astronomers classify M95 as a barred spiral galaxy, but you need a large scope to detect the bar.

Another spiral galaxy from Charles Messier's catalog lies just 3° southeast of the 3rd-magnitude star Theta (θ) Leonis, in Leo's hindquarters. M65 inclines significantly to our line of sight, and so it appears elongated (some four times longer than it is wide). It's not easy to pick out M65's spiral arms, but you should be able to see some irregular structure near the core through a modest-sized telescope.

The brightest of Leo's many galaxies, and arguably the most impressive, is the spiral M66. It lies just east of M65 and belongs to the same galaxy group (as does NGC 3628). M66's spiral arms wrap tightly around the core, although you'll need a 12-inch telescope to make out these arms. In long-exposure photos, pinkish clouds of star formation and the bluish glow of young star clusters dot the spiral arms.

Throughout the universe, it's rare to see isolated galaxies. Most belong to groups containing a few dozen members, or clusters that can harbor thousands of galaxies. The Leo Triplet seen here is one of the more spectacular examples of a small group. The three main members are M66 (upper left), M65 (lower left), and NGC 3628 (right). The three easily fit within a 1° field of view, although you'll have to pump up the magnification to see detail. The group lies about 50 million light-years from Earth.

Although all three members are spiral galaxies, they appear strikingly different to us because they have different orientations. M66 appears closest to face-on, and so shows the most spiral detail. Clouds of nebulosity and bright star clusters pepper its spiral arms. M65 tilts about 15° from edge-on, and so you won't see much detail without a large scope. Finally, NGC 3628 lies edge-on to our line of sight. Its main feature is a dark dust lane that blocks much of the galaxy's light from view.

Scan the Delights in Ursa Major

Ursa Major, the Great Bear, contains none of the sky's 20 brightest stars, yet it stands out more than almost any other constellation. Thank the seven-star Big Dipper, an asterism that knows no equal. The middle star in the Dipper's handle goes by the name Mizar, and it's one of the sky's best double stars. Most of the deep-sky highlights in the Great Bear are galaxies, which include M81, M82, and M101. But you won't want to miss the pretty Owl Nebula (M97).

With the idea of a dipper firmly ingrained in most people's minds, not to mention the prominence of its seven stars, it can be hard sometimes to conceive of anything else in this area of sky. But Ursa Major, the Great Bear, stretches a long way, encompassing more sky than all but two of the 88 constellations. The Dipper's handle represents the Bear's tail, while the bowl forms the main part of the body. Six of the seven stars shine at 2nd magnitude, while the seventh, Delta (δ) Ursae Majoris, glows a magnitude fainter.

You'll need a darker sky to trace the rest of the Bear's outline. A number of fainter stars to the bowl's west (right on the map and photo) creates his neck and head, while several long strings of stars drop south to form the legs. Most of Ursa Major's bright deep-sky objects cluster near the Dipper. Only the intriguing galaxy pair of M81 and M82 lies far away, near the Bear's northern border.

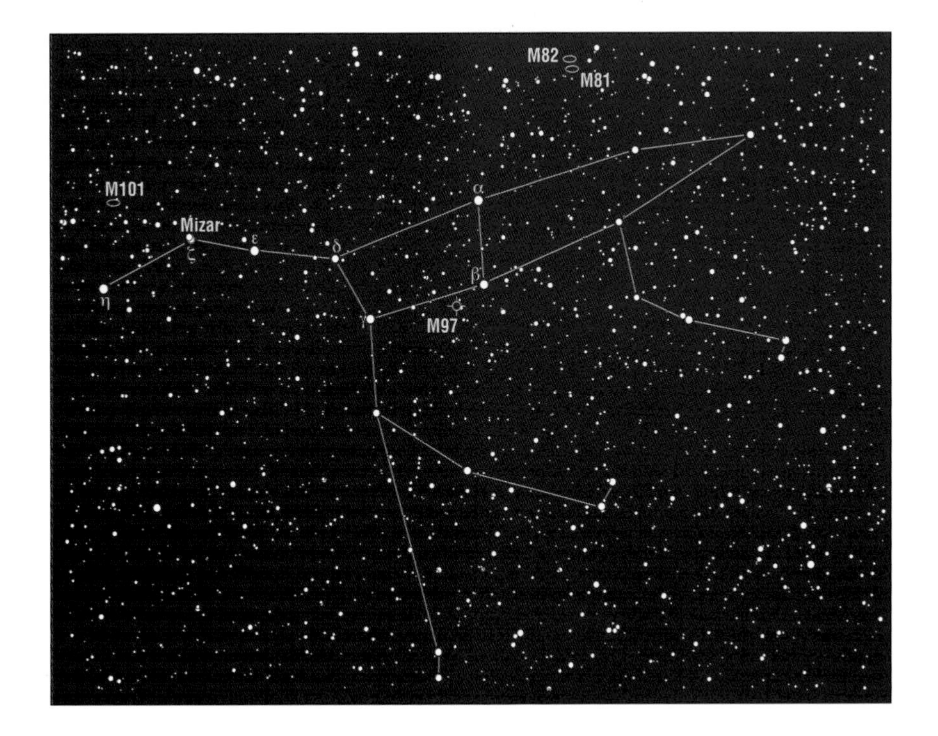

Because the Big Dipper lies within 40° of the North Star, Polaris, it never sets from latitudes higher than 40° north. (Astronomers refer to such stars as *circumpolar.*) Still, Ursa Major has preferred seasons. It passes nearly overhead on spring evenings, but skirts the northern horizon on autumn evenings. You can find the Great Bear well up in the east during the winter months and equally high in the west on summer evenings. The Dipper does dip below the northern horizon for those in the southern United States, but even from there, it's totally gone for at most a few hours a day.

Of the Dipper's stars, only Alpha (α), the star at the end of the bowl, has a distinct yellow-orange color. The other six stars all look blue-white. That shouldn't be surprising, once you realize that five of them—all except Eta (η) at the end of the handle—belong to a single group of stars. Their common origin and similar masses essentially guarantee they'll look alike.

CONTINUED ON NEXT PAGE

Scan the Delights in Ursa Major *(continued)*

If you want to test your eyesight, Ursa Major provides a stellar target. The star at the bend of the Big Dipper's handle, Mizar, has a companion called Alcor. If you can't split these two with your naked eye, it may be time for a new eyeglass prescription. Turn a small telescope on Mizar, and you'll see it also splits into two stars. Giovanni Riccioli discovered Mizar's binary nature in 1650, making it the first known double star.

Just beneath the Big Dipper's bowl, a nocturnal beast swoops silently. The Owl Nebula (M97) gets its name because of two round, dark holes in the nebulosity. The Owl is a planetary nebula, an object that forms when a Sun-like star near the end of its life ejects its outer layers. The Owl has a low surface brightness, so don't magnify the image more than about 100×. And as with most planetary nebulas, an *OIII filter* (doubly ionized oxygen filter) will help reveal detail.

In the northern corner of Ursa Major lies a galaxy group just 12 million light-years away. Spiral galaxy M81 (left) dominates this group. It glows at 7th magnitude, bright enough to see through binoculars, and detail shows up even in small telescopes. M81's slender companion, M82, is a starburst galaxy. M82's core harbors a lot of star-forming regions, each dwarfing our galaxy's massive Orion Nebula. The red glow comes from ejected hydrogen gas.

Spiral galaxy M101 falls into the category of a *grand-design spiral*—one with clearly defined spiral arms and highly organized structure. Even a quick look at this Hubble Space Telescope image shows how apt the designation is. From a distance of some 25 million light-years, Hubble captured dark dust lanes, bright-blue star clusters, and clouds of nebulosity. To create this image, Hubble scientists combined 51 separate exposures into a single mosaic.

To find M101, notice that it makes an equilateral triangle with the two stars at the end of the Dipper's handle. Although M81 is bright overall, glowing at 8th magnitude, its light spreads out enough that you need a large telescope to see much detail. From a dark site, you can pick out the spiral arms quite easily, and even see several of the massive star-forming regions and giant young star clusters.

Go Galaxy-hunting in Virgo

To the naked eye, Virgo the Maiden consists of one bright star (Spica) and a smattering of fainter ones. But Virgo's not about stars. The Maiden hosts the greatest concentration of bright galaxies in the sky. The Virgo Cluster's heart lies near the constellation's northern border, but you can find member galaxies more than 10° away. The standouts include M84, M86, and M87 near the cluster's core, and M61 and M104 well south of there.

Virgo the Maiden stretches a long way, both east-to-west and north-to-south. In fact, it encompasses more area than any other constellation except Hydra, its neighbor to the south. And southern Virgo is the place to begin your introduction to the constellation. There you'll find 1st-magnitude Spica, Virgo's brightest star. It represents an ear of wheat held by a goddess long associated with agriculture. To find Spica, follow the arc of the Big Dipper's handle first to Arcturus in Boötes, then to Spica.

Spica marks the southern corner of a five-sided box that represents Virgo's body. Two lines of stars extending eastward from this box form the Maiden's legs, while westward extensions create her head and arms. The center of the Virgo Cluster, where galaxies crowd closest together, lies near the constellation's northern border. Several cluster members even spill over into Virgo's northern neighbor, Coma Berenices. Most of our highlighted galaxies gather near the cluster's heart, although the beautiful Sombrero Galaxy (M104) lies some 25° away.

Virgo reaches its highest point in the south around 3 a.m. daylight time in mid-March and 11 p.m. in mid-May. From mid-northern latitudes, the constellation's central regions rise less than six hours before these times and set an equal number of hours after. Fortunately, the so-called "Realm of the Galaxies" stands near Virgo's northwestern corner, and so it peaks an hour earlier, reaches a higher altitude, and remains visible nearly an hour longer.

Spica stands out for its blue-white hue. Of all the 1st-magnitude stars visible from the Northern Hemisphere, none appears bluer. The striking color comes about because it has a hot surface, some five times hotter than our yellow Sun. The rest of Virgo's stars will look white without optical aid, because they're too faint to excite our eyes' color receptors.

CONTINUED ON NEXT PAGE

Go Galaxy-hunting in Virgo *(continued)*

M61 ranks among the largest spiral galaxies in the Virgo Cluster. It lies in the cluster's suburbs, nearly 10° south of the jam-packed center. The extra space seems to give this galaxy room to thrive, and the spiral arms display plenty of star formation. Even an 8-inch telescope will show M61's spiral structure. High magnifications will bring out the galaxy's bright core and hint at the expansive star-forming regions.

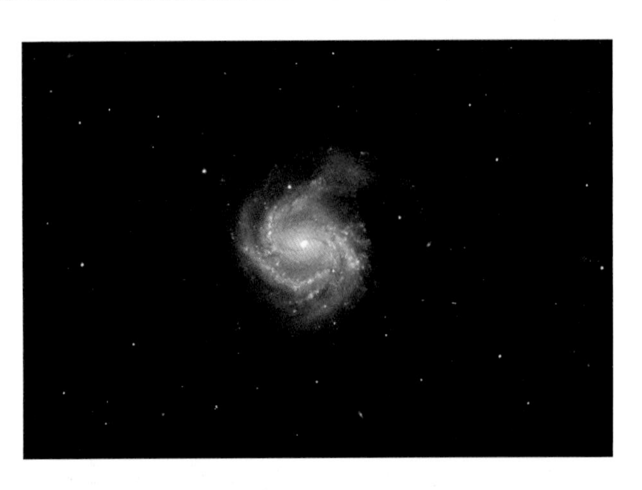

In the world of galaxies, there's big, and then there's BIG. M87 belongs to the latter class. Astronomers estimate this behemoth contains three trillion times the Sun's mass, ten times as much as the Milky Way (which itself is a large galaxy). A supermassive black hole lies hidden at M87's center. Because it is an elliptical galaxy, M87 doesn't show much structure through most backyard telescopes, but you should see its round shape and bright core.

With a dark dust lane reminiscent of a Mexican hat's broad rim, it's no wonder this object is called the Sombrero Galaxy (M104). This spiral galaxy tilts just 6° from edge-on, and so we see its dust lane crossing just south of the galaxy's center. Even a 4-inch telescope will reveal the dust lane, although only near M104's center. Through larger telescopes, you'll be able to trace the dark lane across the galaxy's entire girth.

The Virgo Cluster is the largest gathering of galaxies in the nearby universe. It contains more than 1,000 galaxies that spread across several degrees in Virgo and its northern neighbor, Coma Berenices. The cluster spans millions of light-years and resides some 60 million light-years from Earth. The Virgo Cluster is so big, its gravity pulls on our *Local Group*—the collection of a few dozen galaxies to which the Milky Way belongs—making us part of the so-called "Virgo Supercluster."

This photograph shows only the central 4° of the cluster. The brightest galaxy in the core is M87, which lies at the image's bottom center. Almost as big are M84 and M86, the bright, yellowish balls at lower right. The spiral galaxy near the top is M88 in Coma Berenices. Because most of the bright galaxies here are ellipticals (including M84, M86, and M87) that stopped producing stars long ago, they glow with the yellow-orange hue typical of older stars.

chapter 6

Explore the Summer Sky

In a seemingly cruel twist of fate, the most comfortable observing weather of the year coincides with the shortest nights. Yet summer nights more than compensate for their brevity. Bright stars, gleaming star clusters, and glowing stellar nurseries fill the sky with visual delights.

The dominant star pattern at this time of year goes by the apt name "The Summer Triangle." But the true highlights of the summer sky dot the most spectacular portion of the Milky Way. Whether you view with your naked eye, binoculars, or a telescope, you'll have plenty of sights to keep you satisfied no matter how long the night.

Learn the Summer Constellations

Although it stands highest in the sky on spring evenings, the Big Dipper remains a useful guide throughout the summer. It lies in the northwest as darkness falls. Follow the arc of its handle first to Arcturus in Boötes and then to Spica in Virgo. Summer evenings offer you a last look at these luminaries of spring. Look above Boötes for the semicircle of Corona Borealis and sprawling Hercules. These two bridge the gap between the final spring stars and the first of those traditionally associated with summer.

Above Hercules and nearly overhead on summer evenings, you'll find the three bright stars of the Summer Triangle. This large asterism spans three different constellations. The brightest of the three stars—and the fifth-brightest star in the entire sky—is Vega in the constellation Lyra the Harp (or Lyre). Although it's compact, Lyra stands out quite noticeably. Look for four stars that make a distinct parallelogram hanging off one side of Vega. Just northeast of Vega lies an outstanding double star, Epsilon (ε) Lyrae. Through binoculars, Epsilon splits into two stars. With the greater magnification a telescope provides, each member of this pair splits into two stars. Backyard stargazers have nicknamed Epsilon "the double-double."

East of Vega lies the second member of the Summer Triangle: Deneb in the constellation Cygnus the Swan. Cygnus lies smack in the middle of the Milky Way, and so it can be difficult to trace its pattern under a dark sky. Deneb marks the tail of the Swan, while the star Albireo to the south signifies its head. The swan's wings extend east-west from a bright star immediately south of Deneb. Other people see these stars as the "Northern Cross," with Deneb marking the cross's top. Under a dark sky, you'll see that the Milky Way splits into two parts in Cygnus. This great rift arises because thick clouds of interstellar dust block starlight from beyond.

The Summer Triangle's third member lies farthest south. Altair belongs to the constellation Aquila the Eagle, and appears noticeably brighter than Deneb (but fainter than Vega). A large diamond pattern forms the body of the Eagle, while the raptor's tail stretches south. In the confused avian behavior represented in the summer sky, Aquila flies northward while Cygnus heads south.

Keep heading south from Aquila along the Milky Way, and you'll next run into diminutive Scutum the Shield. Scutum contains no bright star, but does hold the vast Scutum Star Cloud. Millions of distant stars create this cloud, which appears as an especially bright spot in the Milky Way.

South of Scutum lies the heart of the Milky Way: the bright constellation Sagittarius the Archer. The center of our galaxy lies near Sagittarius's western border, and so the constellation includes plenty of star clusters and star-forming nebulas. Don't expect to see an Archer in the constellation's stars, however. Sagittarius's dominant asterism comprises eight stars in the shape of a teapot. In this view, the Milky Way looks like steam rising from the teapot's spout.

Only slightly less prominent than Sagittarius is its western neighbor, Scorpius the Scorpion. Anchored by the 1st-magnitude star Antares, Scorpius does bring to mind a stinging arachnid. Antares marks the Scorpion's heart, while a long, graceful curve of stars represents the body and stinger. It's worth exploring the Milky Way in Scorpius and Sagittarius with binoculars while under a dark sky. Northern observers are at a distinct disadvantage, however, because these constellations never climb high above the southern horizon.

Our final summer constellations lie west and north of Scorpius. Immediately west of the Scorpion (and east of Virgo) is Libra the Scales. Its four most prominent stars form a large diamond shape. A much larger constellation stands north of Scorpius. Ophiuchus the Serpent-bearer possesses several moderately bright stars but none that truly gleam. You'll want to seek a dark sky to trace their arrangement.

As its name implies, Ophiuchus holds another constellation: Serpens the Serpent. Serpens has the distinction of being the only constellation divided into two separate parts. Serpens Caput lies west of Ophiuchus and forms the Serpent's body and head. Serpens Cauda lies east of Ophiuchus and represents the Serpent's tail. The Serpent's meandering string of stars doesn't stand out, but it can make for a nice interlude before diving back into the Milky Way's riches.

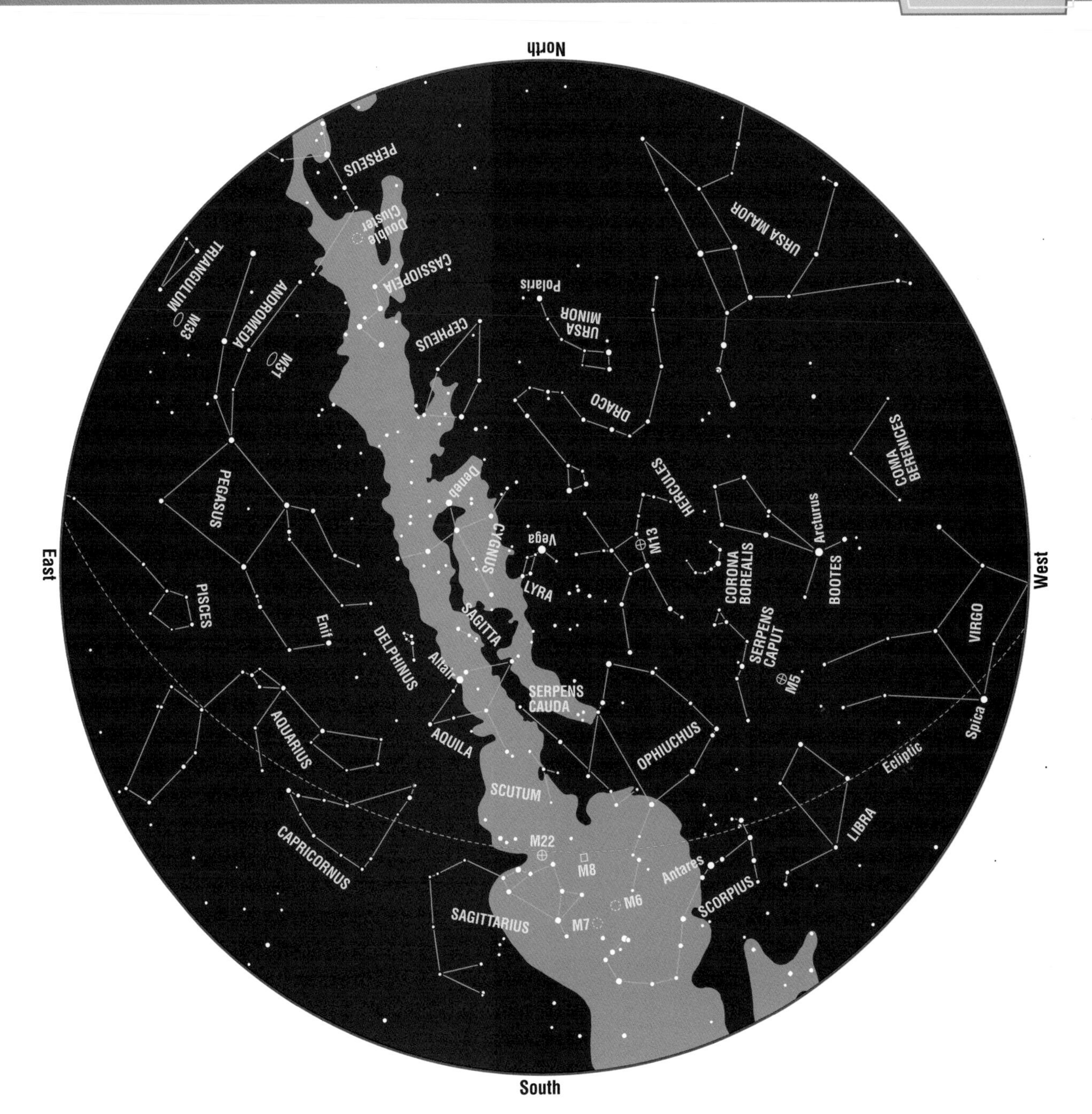

This map shows the entire sky visible from 35° north latitude at approximately 3 a.m. on June 1, 2 a.m. on June 15, 1 a.m. on July 1, midnight on July 15, 11 p.m. on August 1, and 10 p.m. on August 15.

Spy the Glittering Clusters in Scorpius

Scorpius the Scorpion claims one 1st-magnitude star, Antares, and several bright star clusters. Several of these clusters glow bright enough to see with the naked eye, and binoculars or a small telescope reveal them in great detail. The globular clusters M4 and M80 lie near Antares, while the open clusters M6 and M7 stand just north of the Scorpion's stinger. The Bug Nebula (NGC 6302) resides within the Scorpion's body.

With a little imagination, you can trace the shape of a scorpion from the stars of Scorpius. A short line of stars at the western end forms its head. From there, trace a flowing string of stars south and east. First-magnitude Antares marks the Scorpion's heart, and appropriately shines with a reddish color. (The name "Antares" refers to this color. The name comes from the Greek words *anti Ares*, which mean "rival of Mars.") After the star-string bottoms out, it curves back northward and ends at the "Stinger," marked by the star Shaula. In some myths, the Scorpion's sting killed Orion, which is why these two constellations were placed at opposite positions in the sky.

As befits a constellation near the center of the Milky Way, Scorpius features several bright star clusters. M6 and M7 are the two most obvious. They lie north of the Scorpion's stinger and appear to the naked eye under a dark sky. Two more distant clusters, M4 and M80, lie within a few degrees of Antares. A dying star released the gas that forms the Bug Nebula (NGC 6302). You can find this glowing remnant wrapped within the Scorpion's tail.

Scorpius never climbs high in the sky from mid-northern latitudes, and so this is one region where it pays to observe when it's highest in the south. The Scorpion peaks around midnight in mid-June and by 8 p.m. (during twilight) in mid-August. The bulk of the constellation rises about four hours before these times and sets an equal amount after. But the constellation's southernmost stars remain visible for a shorter time and peak at an altitude of less than 10° for those at 40° north latitude.

The most colorful star in Scorpius is 1st-magnitude Antares. It sports a distinctive ruddy hue obvious to anyone who glances in its direction. You should also notice the blue-white color of the slightly fainter star, Shaula. Because of its low altitude from the Northern Hemisphere, this star often twinkles fiercely. This photo also reveals the bright star clouds and dark dust lanes that mark the Milky Way's course through Scorpius.

CONTINUED ON NEXT PAGE

Star cluster M7 spans more than twice the diameter of the full Moon despite a distance of 900 light-years. Its 100-plus stars combine to glow at 3rd magnitude, making the cluster easy to see with the naked eye. Look for M7 about 5° northeast of Shaula. Also look for the cluster's neighbor, M6, 4° to the northwest. M6 lies twice as far from Earth as M7, and so it appears smaller and fainter. Still, it's easy to pick out with the naked eye.

Bugs may not be as dangerous as scorpions, but you wouldn't want to visit the Bug Nebula (NGC 6302) in Scorpius. This planetary nebula—the final gasp of a Sun-like star—features thick clouds of gas and dust ejected by the star. The dying star itself lies hidden within a dusty ring at the upper right of this image. It ranks among the hottest stars known, with a surface temperature of several hundred thousand degrees.

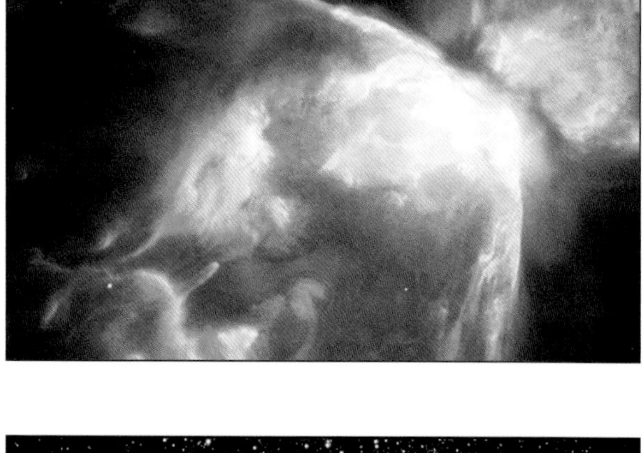

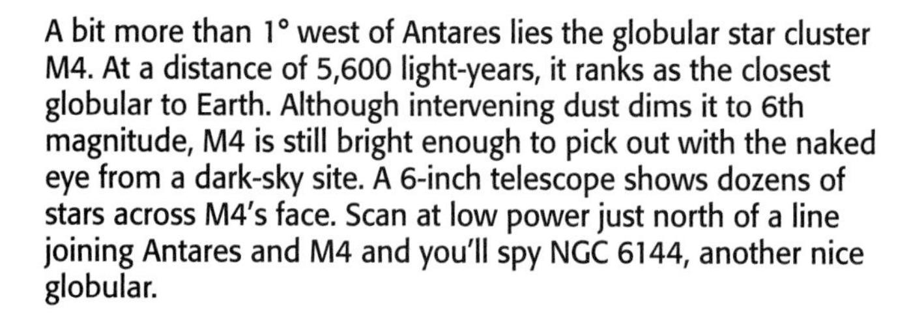

A bit more than 1° west of Antares lies the globular star cluster M4. At a distance of 5,600 light-years, it ranks as the closest globular to Earth. Although intervening dust dims it to 6th magnitude, M4 is still bright enough to pick out with the naked eye from a dark-sky site. A 6-inch telescope shows dozens of stars across M4's face. Scan at low power just north of a line joining Antares and M4 and you'll spy NGC 6144, another nice globular.

Although open star clusters show a lot of variety, most globular star clusters look pretty similar. That's because most globulars pack a few hundred thousand stars into a tightly packed sphere. Most also lie thousands of light-years from Earth, which makes it harder to see distinctions.

The globular seen here, M80, ranks among the densest in our galaxy. It takes the sharp eye of the Hubble Space Telescope to resolve the many stars seen here, and even with Hubble, the central regions blur together into a fiery ball. (An 8-inch backyard telescope will reveal just a handful of M80's stars.) All the cluster's stars formed at the same time, approximately 13 billion years ago. So, they all are more highly evolved than our 4.5-billion-year-old Sun. That's why most of the stars revealed by Hubble have a distinct yellow to orange-red cast. More-massive blue and white stars evolve quickly, and M80's contingent of them perished long ago.

Encounter the Milky Way in Sagittarius

Sagittarius the Archer hosts none of the sky's 20 brightest stars, yet it more than makes up for this lack with a lot of great deep-sky objects. When Charles Messier cataloged the best and brightest of these objects, he saw fit to include 15 from Sagittarius, more than from any other constellation. The variety ranges from splashy star clusters to bright gas clouds—just what you'd expect from the constellation that houses the Milky Way's center.

Sagittarius the Archer shares a lot with its northern cousin, Ursa Major. Just as the Big Dipper asterism dominates the constellation of the Great Bear, Sagittarius is recognized more for its Teapot asterism than for the shape of a bow-and-arrow shooter. The four stars Phi (φ), Sigma (σ), Tau (τ), and Zeta (ζ) Sagittarii form the Teapot's handle; Lambda (λ) represents the lid; and Delta (δ), Gamma (γ), and Epsilon (ε) Sagittarii make the spout. Under a dark sky, the thick clouds of the Milky Way look a bit like steam rising from the spout.

The rest of the Archer trails off to the east and south, but all these stars glow significantly fainter than those in the Teapot. Strangely enough, Sagittarius is not fully human, but rather a centaur (half man and half horse). He represents one of the sky's two centaurs, the other being Centaurus. Most of the Archer's stellar nurseries and open star clusters congregate northwest of the Teapot, where the Milky Way passes. The constellation's globular clusters spread more widely, with several within the Teapot's boundaries and the rest south and east of there.

For Northern Hemisphere observers, Sagittarius suffers from the same low-altitude problem as its western neighbor, Scorpius. The Archer lies highest in the south around 2 a.m. local daylight time in mid-June, and around 10 p.m. in mid-August. The constellation rises and sets approximately four hours before and after those times. Fortunately, most of Sagittarius's top deep-sky objects lie in the northern part of the constellation, placing them higher in the sky for Northern Hemisphere observers.

The bright stars of the Teapot show a pleasing array of colors. Half of the eight stars shine with a blue-white hue, while the other four display yellow to orange colors. This photo highlights the Teapot and the Milky Way band that runs west and north of it. The thick dust clouds at the upper right block our galaxy's very center from view. The bright, reddish cloud at top right is the Lagoon Nebula (M8), visible to the naked eye and a wonder through a small telescope.

CONTINUED ON NEXT PAGE

Encounter the Milky Way in Sagittarius *(continued)*

When you think of Sagittarius, think stars—a lot of stars. Nowhere is this more obvious than in the Small Sagittarius Star Cloud (M24), a sprawling collection of millions of stars. To the naked eye, M24 appears as one of the Milky Way's brightest sections. But you'll want to train binoculars or a low-power telescope at this region. A relative lack of dust in this direction allows us to see stars more than 10,000 light-years from Earth.

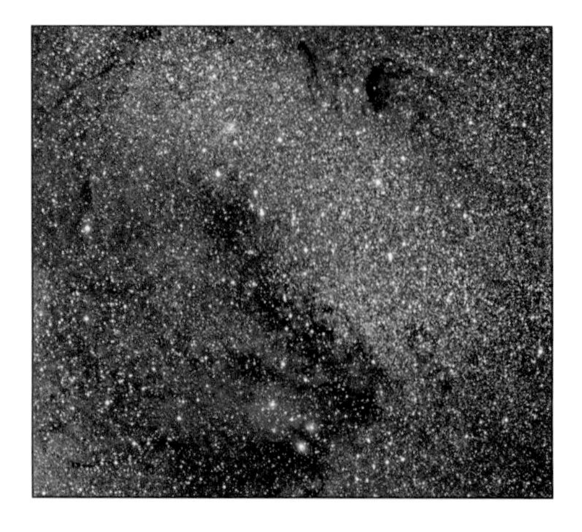

Near the northern edge of Sagittarius, you'll find one of the constellation's signature *emission nebulas*—a glowing cloud of gas and dust where stars are being born. The Omega, or Swan, Nebula (M17) appears as a bright bar through small telescopes. At low power, a short extension curves southward from the bar's western edge. (The extension lies outside the central region shown in this photo.) At higher power, the structure looks like a hook with a nearly black center.

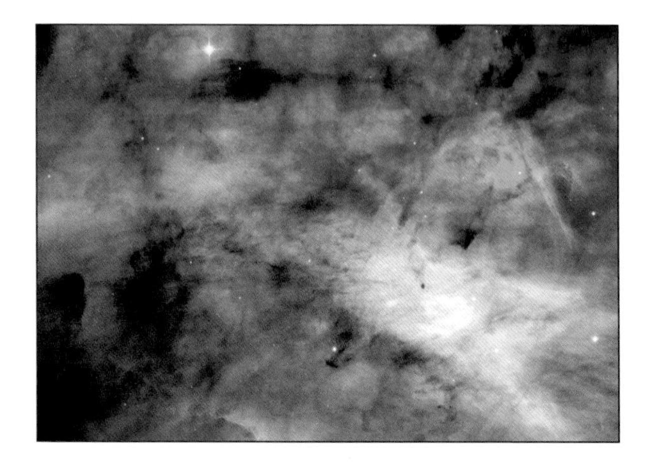

The Trifid Nebula (M20) takes its name from three dust lanes silhouetted against a glowing emission nebula. These dark lanes appear to converge at the nebula's center. Emission nebulas consist largely of hydrogen gas. Ultraviolet radiation from hot, young stars inside the nebula excites the gas to glow with the characteristic reddish hue seen in photos. The blue glow just north of the Trifid comes from light reflecting off dust.

The finest emission nebula within Sagittarius's borders has to be the Lagoon Nebula (M8). This massive region lies just north of the spout of the constellation's Teapot asterism. From a dark site, you can pick out this immense cloud of gas and dust with your naked eye. A small telescope starts to show the Lagoon's full extent, which spans an area larger than the Full Moon. A telescope will also reveal a dark dust lane—the lagoon—cutting the nebula in half.

The force of gravity is causing this cloud to contract slowly, breaking into ever-smaller pieces that eventually will form stars. In fact, plenty of massive stars already have been born here. Look just to the east (left) of the dust lane and you'll see a few dozen newly minted stars embedded in the nebulosity.

Survey Bright Gas Clouds in Cygnus

Cygnus the Swan possesses some of the summer Milky Way's finest treasures. And, unlike its southern cousins Sagittarius and Scorpius, Cygnus passes overhead. Luminous Deneb rules this constellation, whose stars form a distinct cross shape. As with most Milky Way regions, Cygnus features a lot of glowing gas clouds. Among the best are the North America and Crescent emission nebulas, the Veil Nebula supernova remnant, and a "blinking" planetary nebula.

Cygnus the Swan stands astride the Milky Way. Its luminary is 1st-magnitude Deneb, the dimmest of the Summer Triangle's three stars but still the sky's 19th brightest overall. Located some 1,500 light-years from Earth, Deneb ranks as the most distant of the 1st-magnitude stars. It's some 60,000 times more luminous than the Sun. Cygnus provides two obvious patterns—both made from the same stars. The constellation's Swan has Deneb as the tail and the colorful double star Albireo, to the south, as the head. Be sure to view this double at low power, which will show a stark contrast between blue and orange components. The Swan's wings extend east-west (perpendicular to the Milky Way) from 2nd-magnitude Gamma (γ) Cygni. Other people see the same group of stars as the "Northern Cross," with Deneb marking the cross's top.

One of the sky's best emission nebulas lies just east (left) of Deneb. The North America Nebula (NGC 7000) shows up to the naked eye under a dark sky. The Crescent Nebula (NGC 6888) sits near Gamma Cygni, and the huge Veil Nebula sprawls across southern Cygnus. The intriguing "Blinking Planetary" Nebula (NGC 6826) lies near the constellation's northern border.

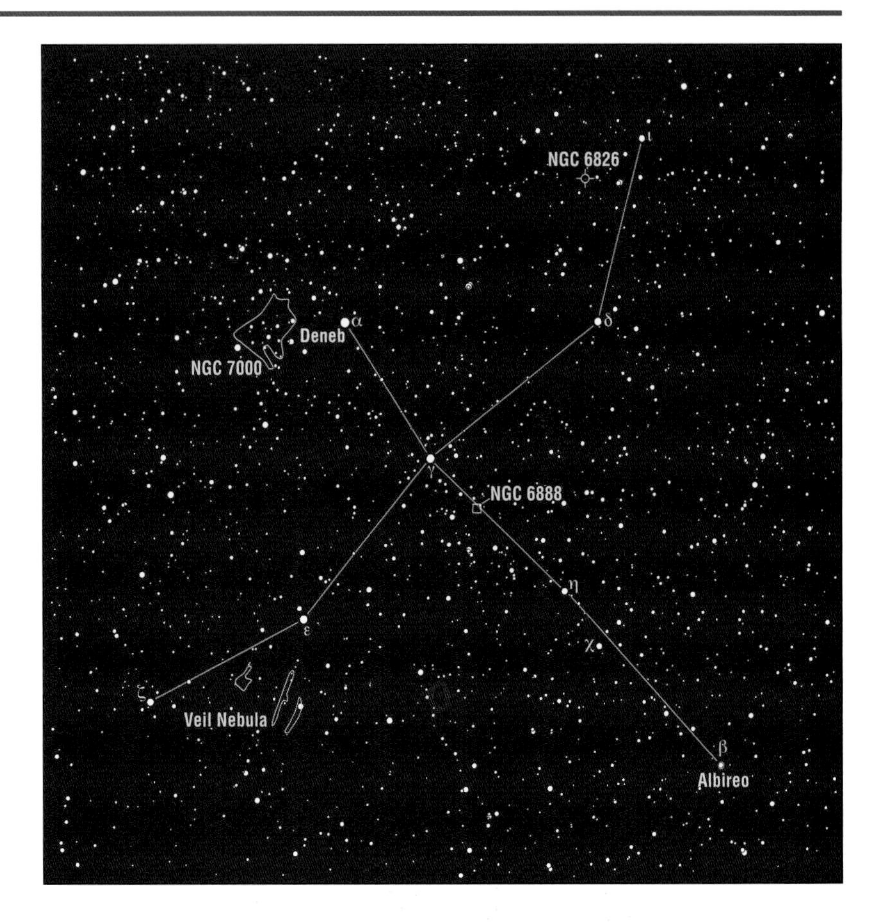

Thanks to its northern location, Cygnus passes overhead for observers at mid-northern latitudes. It reaches this position around 4 a.m. daylight time in mid-June, and about midnight in mid-August. Its northern location also means it stays visible for most of the night. Deneb, for example, remains above the horizon for 20 hours. When Cygnus rises in the northeast, it comes up with its long axis parallel to the ground. As it sets in the northwest, however, the Swan appears to be diving head-first toward the horizon.

Cygnus displays stars of all different colors, although most will look white to the naked eye. The gas clouds to the east (left) of Deneb and completely surrounding centrally located Gamma (γ) Cygni appear reddish only on long-exposure photographs. But the most impressive feature in Cygnus under a dark sky is the broad band of the Milky Way, which is interrupted by a rift of dense interstellar dust.

CONTINUED ON NEXT PAGE

Survey Bright Gas Clouds in Cygnus *(continued)*

A bubble of interstellar gas known as the Crescent Nebula (NGC 6888) lies just 3° southwest of Gamma (γ) Cygni. This cloud formed when the massive blue star seen near the center expelled its outer layers. High-energy radiation from this star causes the cloud to glow. Use low to moderate magnification to see the entire bubble, which glows brightest on its western (right) edge. An OIII filter helps to accentuate the glow.

With a name like the Blinking Planetary Nebula (NGC 6826), you know this object has an intriguing story. The planetary nebula's central star is the reason. Look directly at this star through a small telescope, and you'll see it shining brightly. Glance to the side with averted vision, however, and the star disappears into a haze of nebulosity. The effect vanishes with larger telescopes, which always show the star.

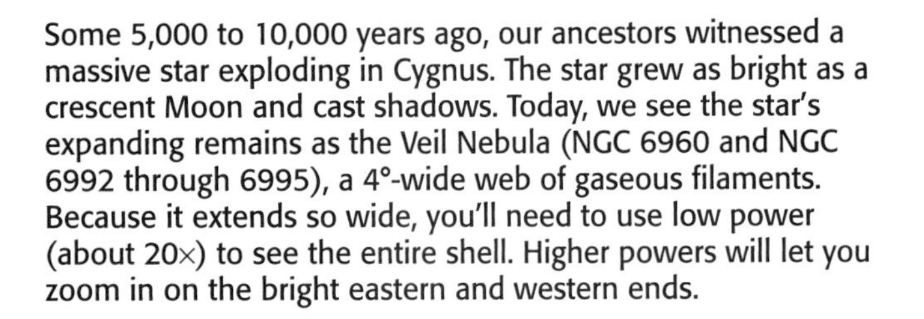

Some 5,000 to 10,000 years ago, our ancestors witnessed a massive star exploding in Cygnus. The star grew as bright as a crescent Moon and cast shadows. Today, we see the star's expanding remains as the Veil Nebula (NGC 6960 and NGC 6992 through 6995), a 4°-wide web of gaseous filaments. Because it extends so wide, you'll need to use low power (about 20×) to see the entire shell. Higher powers will let you zoom in on the bright eastern and western ends.

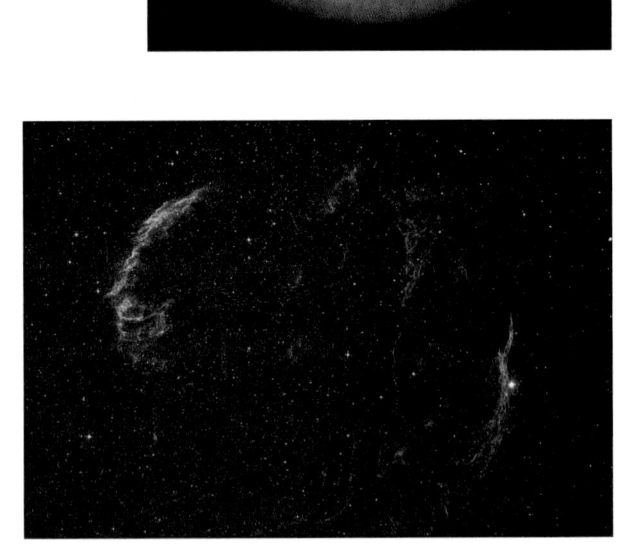

From a dark site, you can spot the North America Nebula (NGC 7000) with just your naked eye. Simply scan some 3° east (left) of Deneb. Better views will come through binoculars or a telescope at low power, which will gather more light and still show the nebula's 2° extent.

The bright nebula really does mimic the shape of North America. Unfortunately, the hazy outline does not look conspicuous. You can improve the contrast markedly with a nebula filter. The continental shape appears most prominent around "Mexico" and "Florida," where bright nebulosity wraps around the dark "Gulf of Mexico." NGC 7000 is a fairly typical star-forming region, where gas and dust are condensing to form a new generation of stars. In the North America Nebula, this process likely will continue for millions of years.

chapter

Explore the Autumn Sky

Perhaps more than any other season, autumn brings to mind change. The shorter days cause tree leaves to change color and then fall. With fewer hours of daylight, the nights also grow longer and cooler. The sky mimics these changes. Sandwiched between the Milky Way treats of summer and the glittering stars of winter, the autumn sky looks bland in comparison.

Don't let this initial impression fool you. Autumn delivers a nice variety of bright star clusters and wispy gas clouds. Yet the season's highlight has to be the Andromeda Galaxy (M31)—the biggest and brightest galaxy visible in the northern sky.

Learn the Autumn Constellations

Autumn is the one season when the Big Dipper takes a vacation. It swings low in the north during the evening hours and even sets for people in the southern United States. Another conspicuous constellation lies on the opposite side of Polaris from the Big Dipper, however. Cassiopeia the Queen has five bright stars arrayed in the shape of the letter "W" or "M," depending on where it stands relative to Polaris. (When it's high in the north during autumn, it looks more like an "M".)

To the east of Cassiopeia lies Perseus the Hero. Although Perseus contains several bright stars, the Hero's shape is not easy to follow. Begin at the constellation's brightest star, centrally located Alpha (α) Persei, and then trace the star pattern from there. Under a dark sky, you'll see a fuzzy patch about one-third of the way from Perseus to Cassiopeia. Binoculars reveal this as the famed "Double Cluster," a side-by-side pair of open star clusters. Perseus's most famous star is Algol, the "Demon Star." It normally shines at 2nd magnitude, but every 2 days, 20 hours, and 49 minutes, it dims by 70 percent. The reason: A faint companion star passes between the brighter component and us, eclipsing much of the bright star's light.

On the opposite side of Cassiopeia from Perseus lies Cepheus the King. Unlike Perseus, the King's stars don't shine brightly but do form a distinct shape. Cepheus looks like a child's drawing of a house, with four stars making a square for its base and a fifth representing the roof's peak.

Looking toward the south on autumn evenings, the most conspicuous group of stars is the Great Square of Pegasus. This set of four 2nd- and 3rd-magnitude stars spans about 15° and stands out because it's big and has few stars visible inside it. (From the suburbs, you'll be lucky to see more than one star inside the Square's borders.) The Square forms the body of Pegasus the Winged Horse, with the head and legs stretching westward. Enif marks the Horse's nose.

Two graceful, curving lines of stars radiate from the Square's northeastern corner. These arcs represent Andromeda the Chained Princess, a constellation best known for harboring the biggest galaxy in our Local Group of galaxies. (The Milky Way ranks second.) The Andromeda Galaxy (M31) appears to the naked eye even from the suburbs, and looks magnificent through binoculars or a small telescope.

Southeast of Andromeda are two small, distinct constellations: Triangulum the Triangle and Aries the Ram. As its name implies, Triangulum consists of three stars in the shape of an isosceles triangle. Its claim to fame is the face-on Pinwheel Galaxy (M33), the Local Group's third-largest galaxy and a fine sight in binoculars or a low-power telescope. Aries the Ram holds three bright stars in a short, crooked line with a fourth star well to their northeast.

Running along the southern and eastern edges of the Great Square are two streams of faint stars that meet at a point. This large, V-shaped constellation goes by the name Pisces the Fish. The southern stream ends in a circular asterism, called the Circlet, directly below the Great Square.

The "V" points directly to one of the more remarkable stars in the sky. Mira varies in brightness by several hundred times over the course of an 11-month period. It typically peaks around 3rd magnitude, bright enough to spot with the naked eye. But more often than not, you'll need binoculars or a telescope to pick up its ruddy glow. Mira stands near the center of the constellation Cetus the Whale, whose head lies northeast of Mira and whose tail waves to the southwest.

To the west of Pisces and Cetus, you'll find two more large, "watery" constellations lacking in bright stars. Capricornus the Sea Goat lies farthest west, and shares a border with the summer constellation Sagittarius. The dozen or so stars that form the shape of Capricornus resemble an arrowhead more than a Sea Goat. Aquarius the Water-bearer has a lot of stars but only one easy-to-recognize pattern. At the constellation's northern edge lies a small, Y-shaped group of four stars called the Water Jar.

Directly south of Aquarius and not too far above the horizon, an isolated bright star will draw your attention. This is 1st-magnitude Fomalhaut, the brightest star in the otherwise drab constellation Piscis Austrinus, the Southern Fish. Of the 25 brightest stars in the sky, Fomalhaut is the lone representative from the traditional autumn constellations. Under a reasonably dark sky, you should see a zigzag line of faint stars that runs from the Water Jar asterism to Piscis Austrinus. These stars represent water droplets flowing into the Southern Fish's mouth.

You may have noticed a theme among the last several constellations—all have a connection to the sea. This vast celestial ocean recalls the time thousands of years ago when the Sun crossed this region during the rainy season. Ancient stargazers forged a link between land and sky that still echoes in the 21st century.

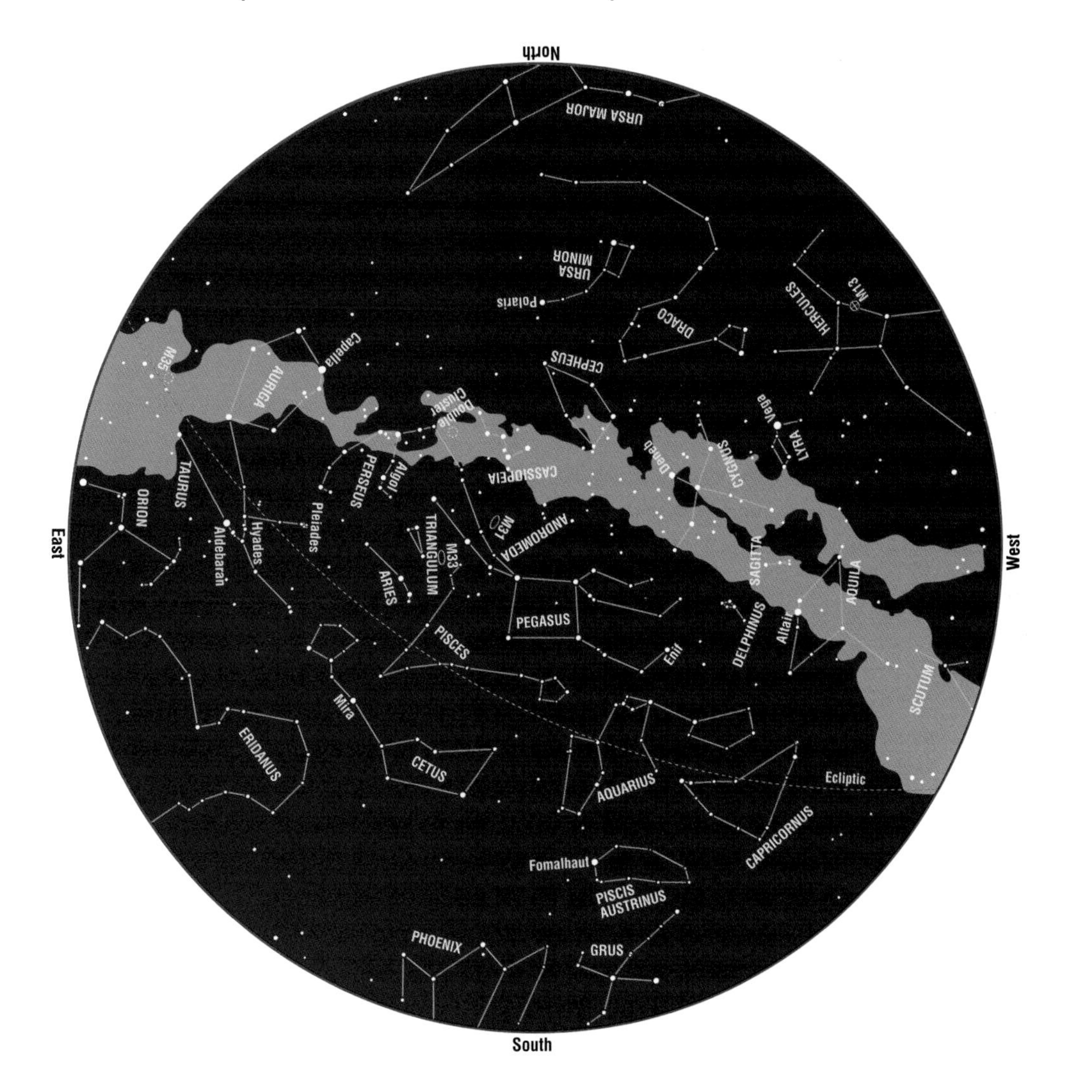

This map shows the entire sky visible from 35° north latitude at approximately 2 a.m. on September 1, 1 a.m. on September 15, midnight on October 1, 11 p.m. on October 15, 10 p.m. on November 1, and 8 p.m. on November 15.

Detect a King's Fortune in Cepheus

Cepheus the King resides along the northern fringe of the Milky Way. Although it claims no star brighter than 2nd magnitude, it does hold several interesting deep-sky objects. The constellation's southern section features the sprawling emission nebula IC 1396 as well as Delta Cephei, a *variable star* whose brightness changes regularly. Farther north, you'll find the superb planetary nebula NGC 40 and, not far from Polaris, the ancient star cluster NGC 188.

When you think of a residence fit for a king, most likely you picture a grand palace or a castle. So when you scour the sky for the celestial king, be prepared for a surprise. Cepheus lives in a humble abode. Connect the brightest stars in this constellation and you won't see a castle, but rather a five-sided figure that resembles a child's drawing of a house with a steep roof. Its connection with royalty may date back some 20,000 years, when Earth's axis pointed toward this constellation. The home of the pole star certainly would merit recognition from ancient skywatchers.

With the Milky Way traversing the southern part of Cepheus, most of the constellation's deep-sky objects congregate there. The glowing gases that make up IC 1396 almost cross the border into Cygnus. And the important star Delta (δ) Cephei lies within a couple of degrees of neighboring Lacerta. The bright planetary nebula NGC 40 breaks this mold from its position in east-central Cepheus. You'll have to search near Polaris to glimpse one of the oldest open star clusters known: NGC 188.

With the exception of Ursa Minor and its asterism, the Little Dipper, no prominent group of stars lies closer to the north celestial pole than Cepheus. This places it in the sky continuously for anyone north of about 30° north latitude. It appears highest in the north and at its best on autumn evenings, however, so that's the best time to view it. Remember that the peak of the roof points north (toward Polaris), which will help you to locate it at other times.

Cepheus has a smattering of blue-white, white, yellow, and orange stars, but the one that stands out most has a distinct reddish hue. Mu (μ) Cephei lies at the northern end of the emission nebula IC 1396. It ranks among the reddest of all naked-eye stars, but it appears more colorful with some optical aid. The star's color so impressed 18th-century observer Sir William Herschel that he dubbed it the Garnet Star.

CONTINUED ON NEXT PAGE

Detect a King's Fortune in Cepheus *(continued)*

Delta (δ) Cephei is worth a king's ransom to astronomers. In 1784, John Goodricke discovered that the brightness of this naked-eye star varies by a factor of nearly two during a 5.4-day period. (You can track the change by comparing it with nearby stars.) Similar stars soon turned up, and scientists realized their periods foretell their actual brightnesses. So, by measuring one's period, you can calculate its distance. Astronomers now refer to such stars as *Cepheid variables*, in honor of Delta.

Located just 4° from Polaris, the star cluster NGC 188 ranks among the sky's northernmost deep-sky objects. A modest-sized telescope reveals about 50 stars packed into a circle whose diameter is nearly half that of the Moon. But NGC 188 stands out mostly for its age. Astronomers estimate that it formed about 7 billion years ago, making it far older than typical open clusters like the Pleiades in Taurus.

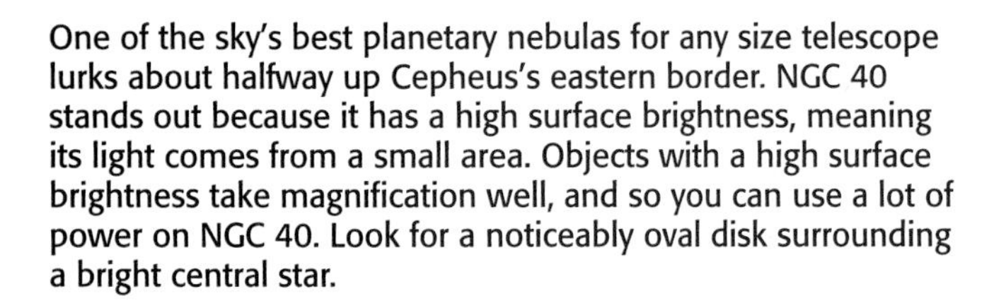

One of the sky's best planetary nebulas for any size telescope lurks about halfway up Cepheus's eastern border. NGC 40 stands out because it has a high surface brightness, meaning its light comes from a small area. Objects with a high surface brightness take magnification well, and so you can use a lot of power on NGC 40. Look for a noticeably oval disk surrounding a bright central star.

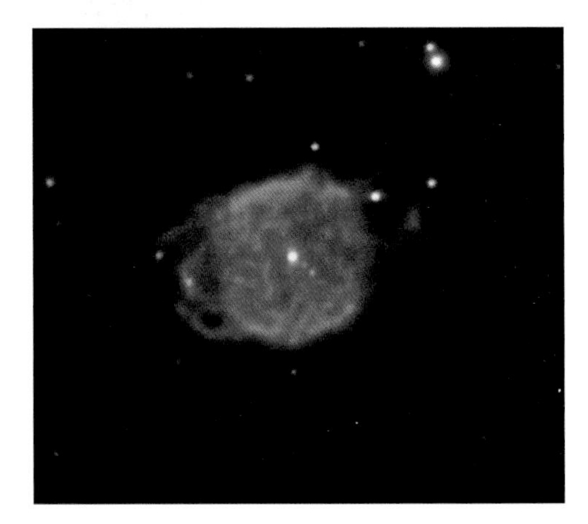

Herschel's Garnet Star serves as a guide to the vast emission nebula cataloged as IC 1396. The red supergiant star—one of the largest stars known in our galaxy—lies at the nebula's northern edge. The gaseous region extends more than 2°, some four times the Moon's diameter. And this is despite its distance of some 2,600 light-years. A small telescope shows IC 1396 as a misty patch, but a nebular filter and larger apertures start to reveal some of the region's darker streaks of nebulosity.

At the heart of the nebula, you'll find a naked-eye star cluster known as Trumpler 37. It packs more than 50 stars into a region about 1° across. At magnifications of 150× or higher, the cluster's center resolves into a triple star, one shining at 6th magnitude and the other two at 8th magnitude. High-energy radiation from these stars causes the nebula to glow. Trumpler 37 appears to be about 3 million years old. That makes it a baby compared with NGC 188, its ancient companion at the opposite end of Cepheus.

Observe Milky Way Riches in Cassiopeia

Cassiopeia the Queen reigns over the late autumn and early winter Milky Way. As such, this conspicuous constellation sees her fair share of bright star clusters and glowing emission nebulas, which span the Queen from one end to the other. Surprisingly, the Milky Way's density in Cassiopeia is low enough that a few faint background galaxies shine through. Most of these belong to our Local Group.

Five 2nd- and 3rd-magnitude stars make up Cassiopeia's prominent form. These stars—Alpha (α), Beta (β), Gamma (γ), Delta (δ), and Epsilon (ε) Cassiopeiae—create a zigzag line that looks just like the letter "W" or "M," depending on the season and the time of night. On autumn evenings, when Cassiopeia stands high in the north and above Polaris, it looks like an "M." When it scrapes the northern horizon on spring evenings, it takes the shape of a "W."

Cassiopeia the Queen and next-door neighbor Cepheus the King are the only husband-and-wife team in the heavens. Two of our deep-sky highlights—the open cluster M52 and the Bubble Nebula (NGC 7635)—lie next to each other on the border with Cepheus. Open cluster NGC 457 and emission nebula NGC 281 reside in central Cassiopeia, slightly south of the letter shape.

Like Cepheus and Ursa Major, Cassiopeia remains visible most or all of the night for people at mid-northern latitudes. If you live north of 35° north, the constellation never sets. It does swing low on spring evenings, however. You can always find Cassiopeia by looking on the opposite side of Polaris from the Big Dipper. The Queen is hardest to recognize in winter and spring, when the "W" or "M" shape lies on its side and appears less evident.

Most of Cassiopeia's bright stars glow with a white or blue-white color. The one exception: Alpha (α) Cassiopeiae, which sports a distinctly ruddy hue. This photo shows north up (as most of our photos do). If you look carefully near the left-center of the photo, you should see two star groups that nearly overlap. This is the spectacular Double Cluster, which lies in Perseus, just across its border with Cassiopeia.

CONTINUED ON NEXT PAGE

Observe Milky Way Riches in Cassiopeia *(continued)*

It usually takes some effort for a star cluster to stand out from a rich Milky Way background. NGC 457 accomplishes this feat with the help of a 5th-magnitude star, Phi (ϕ) Cassiopeiae, that sits on its edge. The cluster itself contains dozens of stars visible through a small telescope. Some observers refer to NGC 457 as "The Owl" because two stellar "wings" appear to emanate from the cluster's center, one to the east and one to the west.

A gaseous glow lies just 2° east of Alpha (α) Cassiopeiae. Images reveal a dark dust lane cutting through the nebula, which is cataloged as NGC 281. The intriguing shape led observers to christen it the "Pac Man Nebula," a reference that seems as dated now as a rotary telephone. The dark gash appears more prominent when viewed through a nebula filter, which helps show the faint nebulosity surrounding it.

The easiest way to find the open star cluster M52 is to draw a line from Alpha (α) to Beta (β) Cassiopeiae and then extend it an equal length. The combined light of M52's nearly 200 stars brings it close to naked-eye visibility, although it falls just short. Binoculars or a small telescope easily show the cluster against the Milky Way backdrop. Through a 6-inch telescope, you can expect to see at least 50 of the cluster's stars.

In a constellation that features somewhat fanciful visions of an Owl and a Pac Man, it's nice to see an object with a can't-miss nickname. One look at the Bubble Nebula (NGC 7635) and you'll know exactly why it got this tag. The Bubble lies in far western Cassiopeia, just half a degree southwest of the open cluster M52.

Stellar winds from an 8th-magnitude star inside the bubble sculpt the surrounding gas into the shape we see, and the star's radiation triggers the gas to glow. The bubble spans approximately 10 light-years and lies more than 1,000 light-years from Earth. The bright star shows up in any telescope, but you'll need an 8-inch telescope to see much nebulosity. A nebula filter improves the view noticeably.

Spot a Grand Galaxy in Andromeda

Andromeda the Chained Princess displays several 2nd-magnitude stars. But stars are the last thing on most observers' minds when they get here. This constellation holds the Andromeda Galaxy, the biggest and brightest galaxy in the northern sky. It also contains a few small companion galaxies, a bluish planetary nebula, and one of the sky's best edge-on galaxies.

Two curving rows of stars form Andromeda's basic shape. The rows meet at the star Alpha (α) Andromedae, which also forms the northeastern corner of the Great Square of Pegasus. Although Alpha completes the Square, it belongs officially to Andromeda. The great Andromeda Galaxy (M31) lies slightly above the northern row of stars and appears as a hazy patch to the naked eye even from the suburbs. M31 has two small companions you'll need a telescope to view. The planetary nebula known as the Blue Snowball (NGC 7662) lies in far western Andromeda, on the outskirts of the Milky Way. And NGC 891, an edge-on spiral galaxy, dwells in the constellation's opposite corner. It stands just a few degrees east of the striking double star Gamma (γ) Andromedae.

Andromeda the Chained Princess was the daughter of Cepheus the King and Cassiopeia the Queen. All three—along with two other autumn constellations—take part in one of the greatest Greek myths. It seems Cassiopeia was a vain woman who thought herself more beautiful than the sea nymphs. Not impressed, the nymphs decided to get revenge. They asked Poseidon, god of the sea, for help. He unleashed Cetus the Whale to terrorize the kingdom. An oracle told Cepheus their only solution was to sacrifice Andromeda by chaining her to a rock on the coast. Enter Perseus the Hero. Promised Andromeda's hand in marriage if he saved her, Perseus killed Medusa and then used her hideous head to turn Cetus to stone.

Andromeda passes nearly overhead for observers at mid-northern latitudes. It reaches this position around 2 a.m. daylight time in mid-September, and about 9 p.m. standard time in mid-November. The constellation rises about nine hours before this peak and sets an equal amount of time after. Pegasus's Great Square serves as a convenient guide for finding Andromeda. The two rise at about the same time, and Andromeda spends the rest of the night trying to catch up to the Winged Horse.

The two rows of Andromeda's stars display different colors. The northern row features mostly blue-white or white stars, while the southern row consists of yellow to reddish suns. Be sure to view Gamma (γ) Andromedae, the star at the end of the southern row, through a telescope. You'll see an orange primary star shining at 2nd magnitude and a nearby blue secondary glowing at 5th magnitude. The Andromeda Galaxy (M31) shows up on the photo just above Mu (μ) Andromedae.

CONTINUED ON NEXT PAGE

Spot a Grand Galaxy in Andromeda *(continued)*

The second-best galaxy in Andromeda would rank number one in a lot of other constellations. NGC 891 is a spiral galaxy oriented nearly edge-on (it inclines just 1.4° to our line of sight). The galaxy appears four times as long as it is wide. A 10-inch telescope shows a narrow streak of light bisected by a dark dust lane. Dozens of foreground stars belonging to our galaxy sprinkle the field, which adds to the visual impact.

If you seek color in deep-sky objects, look no further than the Blue Snowball (NGC 7662). Almost everyone who observes this planetary nebula at low power sees some shade of blue emanating from the expelled gases of a dying Sun-like star. In small telescopes, the gaseous disk appears tiny and evenly illuminated. Larger scopes start to bring out some of the details visible in the photo, including the bright inner ring and fainter gas shell.

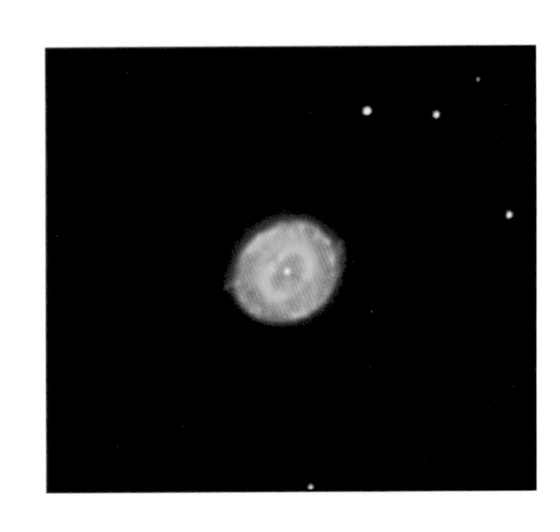

Dwarf galaxies—typically collections of several million stars—account for the lion's share of all galaxies in the universe. But put one of these small entities more than a few million light-years away, and you need a big telescope to see it. A look at the Andromeda Galaxy (M31) with a small scope will reveal two dwarf galaxies. NGC 205 (seen here) appears three times larger but noticeably fainter than M32. Both companions appear in the same low-power field as M31.

You don't need any optical aid to spot the Andromeda Galaxy (M31). You can find it even from a suburban backyard, where the galaxy appears 1° long and a third as wide. From a dark site, these dimensions triple. Any telescope will deliver superb views of M31. A small telescope at low power will show the entire galaxy, which spans more than 3°—that's six Full Moons stacked side by side. A 6-inch telescope reveals a prominent dust lane, and a second one shows up under a dark sky.

With larger telescopes, which can't show the whole galaxy in one field, you'll need to concentrate on specific features. You can spy individual star clusters, trace dark dust lanes, and view the companion galaxies, M32 and NGC 205. It's hard to get bored with an object like this. At a distance of roughly 2.5 million light-years, M31 shows details we can only dream about in more distant galaxies.

chapter

Observe the Sun and Moon

The two brightest objects in the sky provide a unique set of observing challenges and rewards. For the Sun, "challenge" is the key word—even a brief glance without proper eye protection can permanently damage your sight. With safe viewing techniques, however, the Sun offers detail impossible to see on any other star.

The Moon also provides endless hours of enjoyment. Although it always keeps the same face pointing toward us during its monthly trek across the sky, the Moon nevertheless manages to look different every night. Through a backyard telescope, it tells a tale of solar system history you can't see anywhere else.

The Sun and Moon play preeminent roles in two spectacular sky events. Whenever the Sun, Moon, and Earth align precisely, we witness either a solar or lunar eclipse. Given the chance, you should seek out these dramatic affairs.

How to View the Sun Safely

Most people would rank eyesight as their most valuable sense. And that certainly holds true for backyard observers. Yet observing the Sun poses a significant danger to eyesight. The human eye acts like a lens, focusing light on the retina. And the Sun is so bright that it can burn the retina's delicate tissue quickly. The only way to view our daytime star safely is by projecting its image or viewing it through a proper solar filter.

Even without optical aid, the Sun can burn your retina in seconds. And looking through binoculars or a telescope can blind you in a fraction of a second. Don't use any filter or device not specifically designed for viewing the Sun. Although these may block visible light, they let through dangerous amounts of infrared radiation. And always place a safe filter in *front* of the binoculars or telescope—you don't want concentrated sunlight hitting the filter.

WARNING: Viewing the Sun without proper protection risks serious eye damage and possible blindness.

An easy way to view the Sun is to project its image. A homemade pinhole "camera" provides a simple means of projection. Take a stiff piece of cardboard and cut a square hole in it. Then, tape a piece of aluminum foil over the hole and use a straight pin to poke a hole through it. Now, with your back to the Sun, let sunlight pass through the pinhole and onto a second piece of white cardboard. The projected Sun's image grows larger but dimmer the farther apart you hold the pieces.

A safe and inexpensive way to view the Sun with your naked eye is through a #14 welder's glass. These come in 2-by-4-inch rectangular pieces and cost a few dollars at well-stocked welding-supply stores. Be sure you get #14 welder's glass—the more common #12 grade doesn't block enough light. Welder's filters, which render the Sun a greenish hue, let you see *sunspots*—relatively cool and dark regions of the Sun's atmosphere—the size of Earth and bigger.

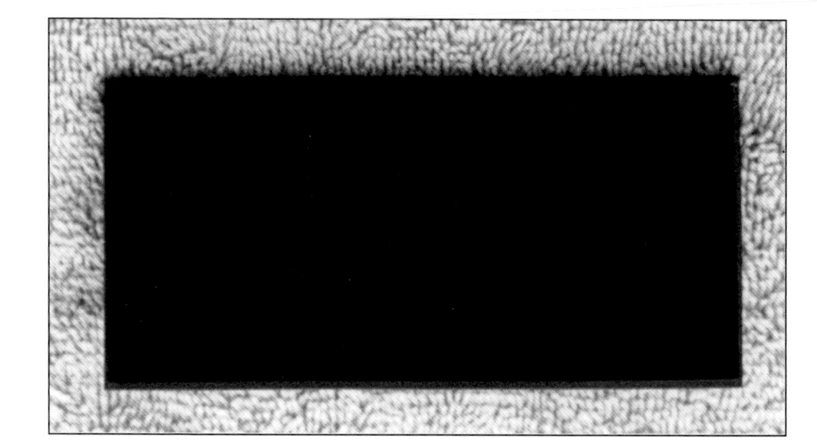

If you want to view the Sun directly through a telescope, you need a filter that fits snugly over the scope's front end. Manufacturers make such filters out of either metal-coated glass or Mylar. (This is not the same Mylar used in "space blankets" and other household or outdoor products. Don't try these as solar filters.) Several companies make solar filters, which you can buy from most large astronomy dealers.

Most solar filters show you the Sun in white light, which emphasizes features on the visible "surface," such as sunspots. *Hydrogen-alpha filters* target a certain wavelength emitted by hydrogen gas and reveal details higher in the solar atmosphere. With these filters, you can see prominences above the Sun's *limb* (the edge of its disk), features that are otherwise invisible except during total solar eclipses. Hydrogen-alpha filters can be expensive, however, costing thousands of dollars.

See the Sun's Changing Face

The Sun is a gigantic ball of gas. It would take more than 100 Earths placed side-by-side to equal the Sun's diameter, and more than a million Earths to fill its volume. It's so big and so close, we can view details on its "surface" that are invisible on any other star. (To see these details, be sure to use a safe solar filter.) Dark sunspots are the most obvious features, but with the right filter, you can see a lot more.

You wouldn't know it by looking at one, but sunspots actually shine brightly. They appear dark only because they pale in comparison to the rest of the Sun's surface. The temperature of the surface—the level from which most of the Sun's light escapes—averages nearly 10,000° Fahrenheit. A sunspot glows at least 2,000° cooler. In a typical sunspot, you'll see a darker central region, called the *umbra,* and a lighter surrounding region, known as the *penumbra.*

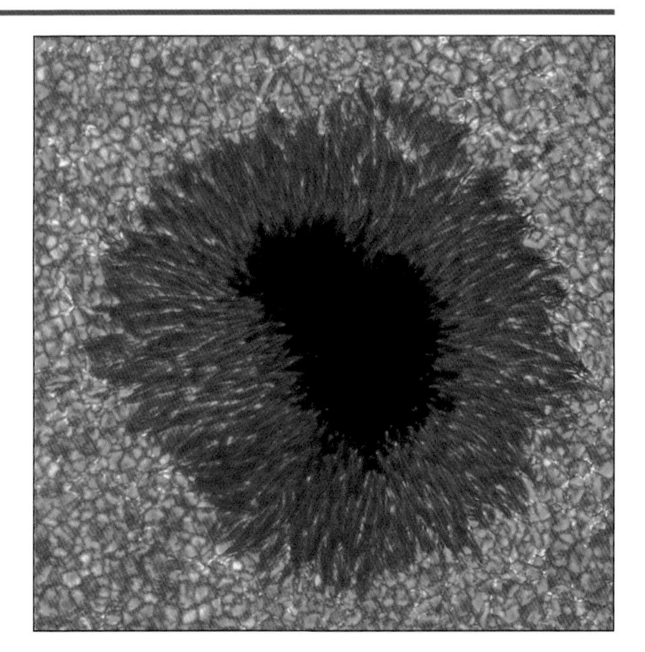

The number of sunspots varies with a period of about 11 years. At the cycle's peak, a lot of large sunspots pock the Sun's surface. At minimum, you can search long and hard without seeing any spots. The last cycle peaked in the year 2000 and reached a minimum in 2008. Astronomers expect the next cycle to peak in 2010 or 2011.

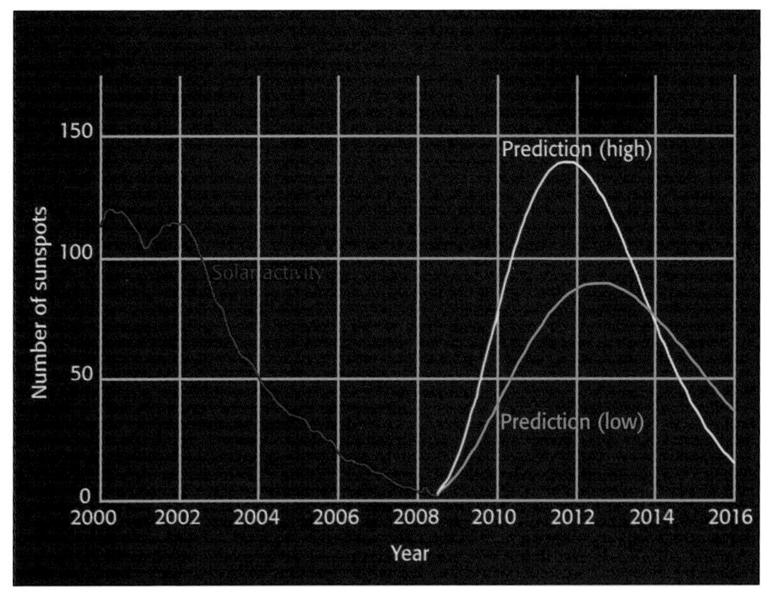

Sunspots start as barely perceptible blemishes on the Sun's surface. Some fade away in a day or two, while others grow huge. The largest sunspots span areas more than ten times the size of Earth. Such spots typically appear in complex groups and last several weeks. The Sun takes about a month to rotate on its axis, so these big groups often survive long enough to pass across the Sun's face more than once.

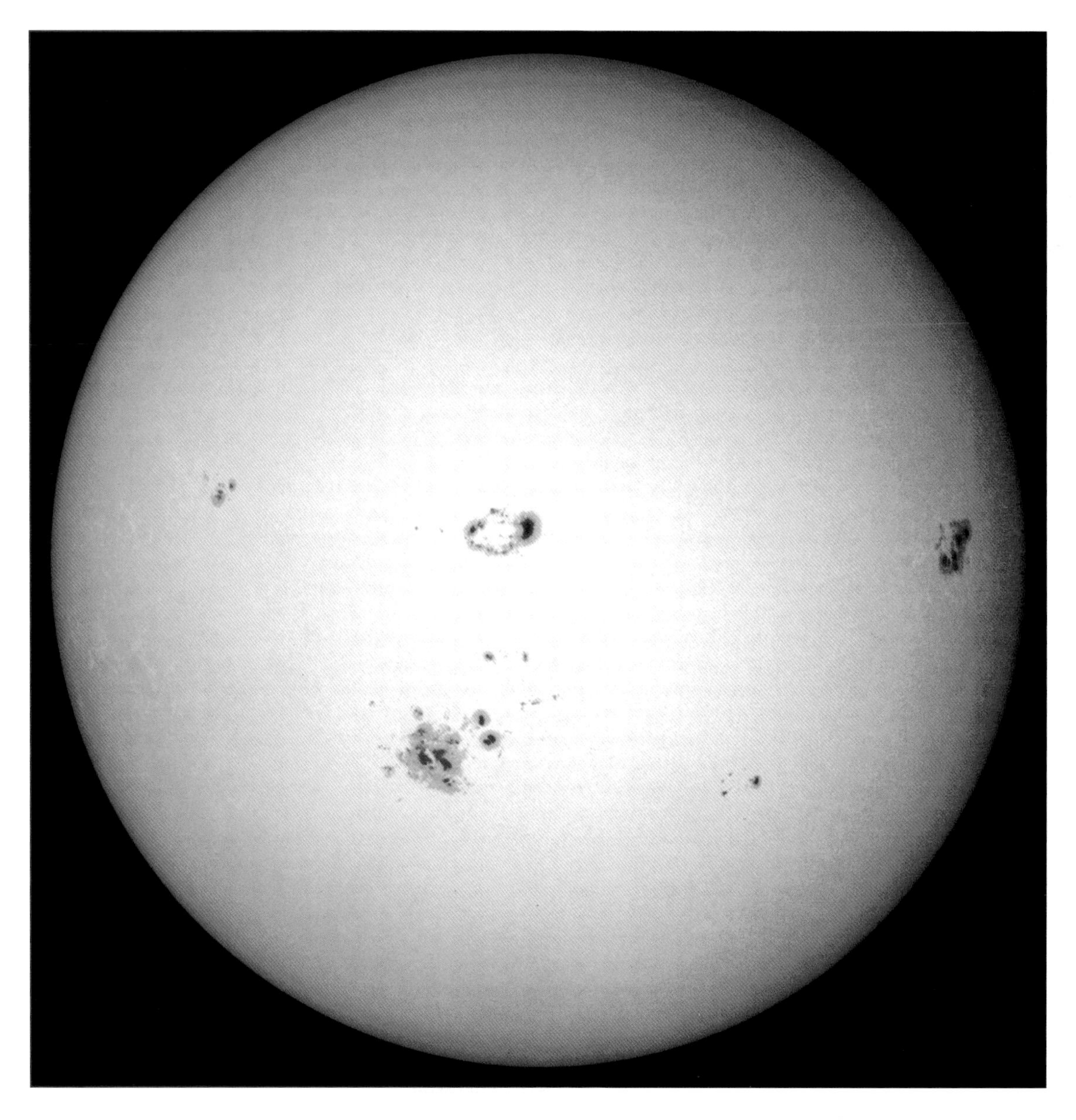

CONTINUED ON NEXT PAGE

See the Sun's Changing Face *(continued)*

Hydrogen-alpha filters transmit just one color of sunlight—that produced by hydrogen gas at a wavelength of 656.3 nanometers. (This same wavelength gives most emission nebulas their characteristic reddish glow.) This color comes predominantly from the Sun's *chromosphere* (the layer just above the visible-light *photosphere*), and so gives a unique perspective on solar activity. In this view, we see a brightly glowing active region.

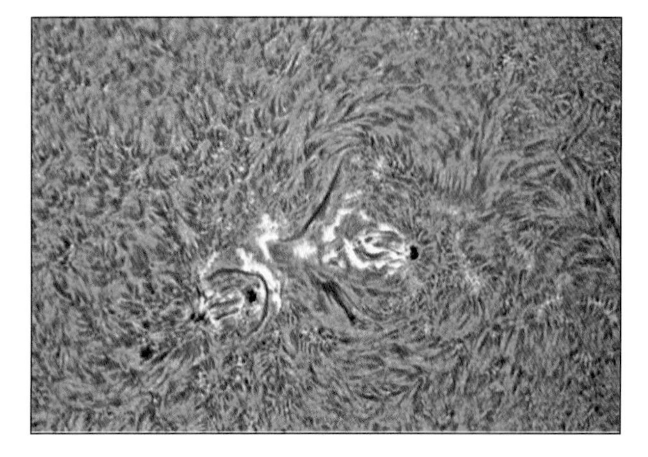

Sunspots take on a different character when viewed through a hydrogen-alpha filter. We still see the dark umbra and the lighter penumbra surrounding it, but we also see gently curving, dark contours. These curves mark the locations of the Sun's magnetic field lines, along which ionized gas flows. The magnetic fields slow the movement of gas near the Sun's surface, causing it to cool and thus appear darker.

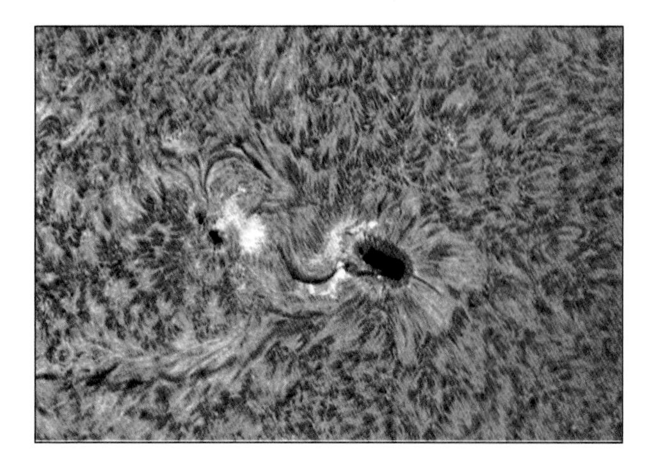

Hydrogen-alpha filters also reveal *solar prominences,* loops of relatively cool, dense gas held in the Sun's chromosphere by magnetic forces. When seen against the blackness of space, prominences appear bright (as in this photo). But when viewed against the brilliant solar disk, they appear as dark arcs known as *filaments*. Without a hydrogen-alpha filter, prominences show up only during total solar eclipses.

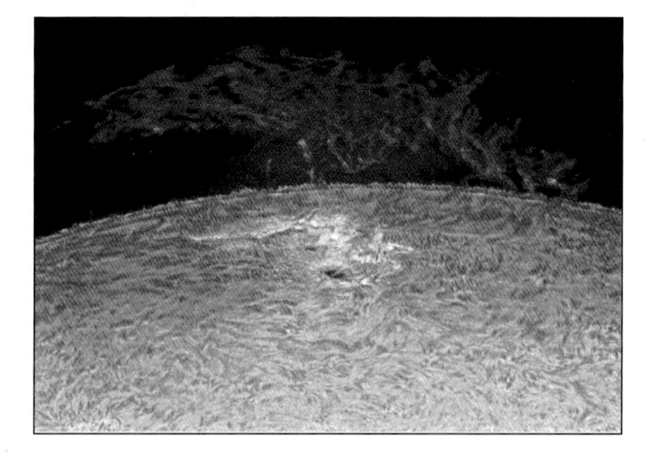

Dark sunspots and bright active regions dominate the Sun's disk in this hydrogen-alpha image, taken during a period of heightened solar activity. Around the limb, a dozen or so prominences leap above the Sun's surface. Look carefully right along the solar limb, and you should see the thin, reddish glow of the Sun's chromosphere. This layer of the solar atmosphere lies just above the photosphere, where the Sun radiates most of its visible light. The chromosphere appears thin because it's only several thousand miles thick, compared with the Sun's diameter of about 865,000 miles.

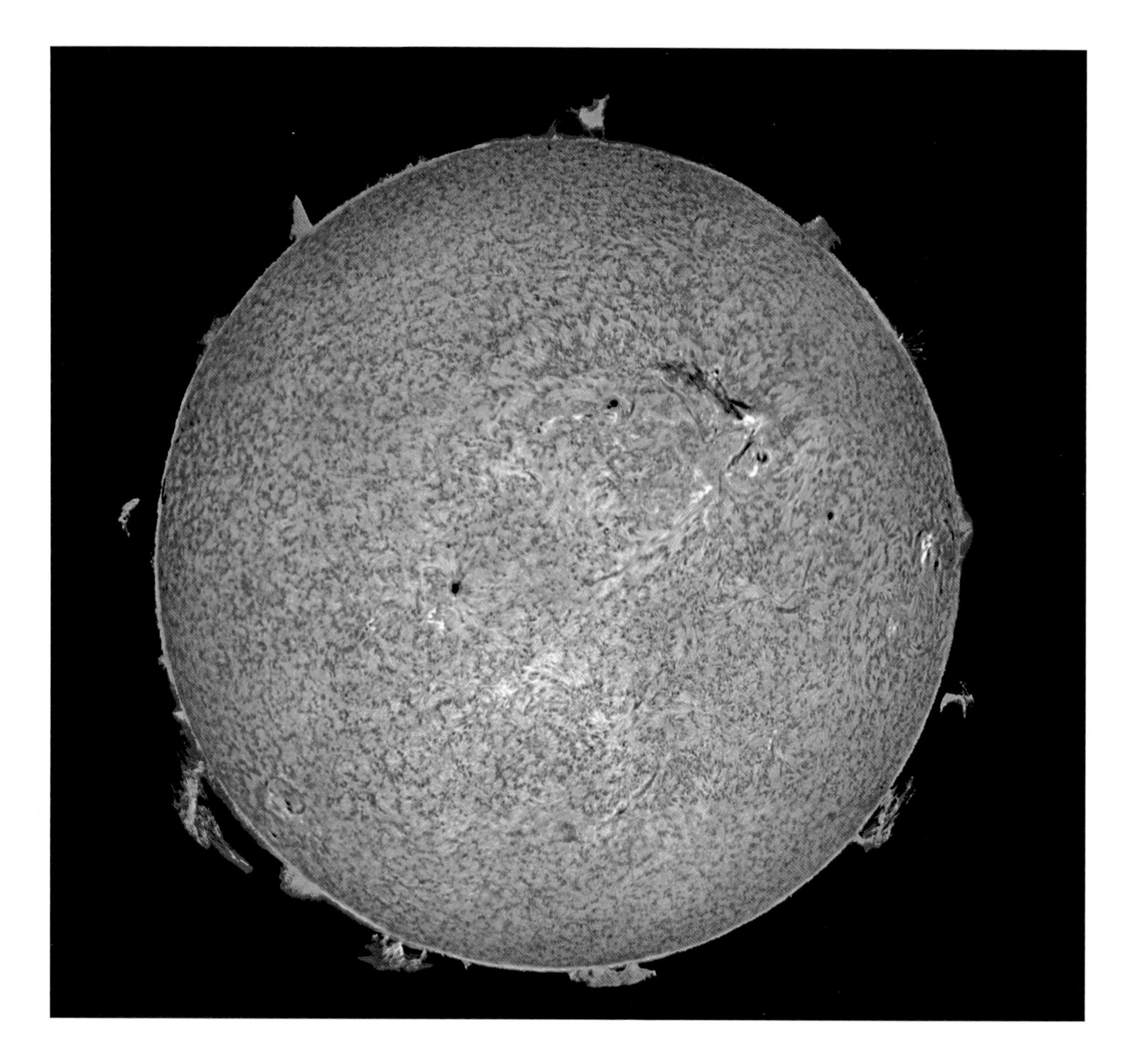

View the Lunar Surface

Even people with no interest in astronomy recognize the Moon. It's so bright you can see it clearly from the city. And it's so big you can make out broad details with your naked eye. Binoculars and small telescopes reveal a lot more. The Moon makes a tempting target for observers—and a rewarding one. Most lunar features show up best where the interplay of light and shadow is strongest, along the "terminator," which divides day and night on the Moon.

The Moon shines by reflecting sunlight. Our celestial companion takes 27.3 days to revolve around Earth. Because our planet revolves around the Sun, however, it takes longer for the Sun, Moon, and Earth to return to the same alignment. This 29.5-day period marks the cycle of lunar phases.

The lunar month begins at New Moon, when the Moon lies between the Sun and Earth. At this time, the Moon's lit hemisphere faces away from us, and we can't see it. The Moon then moves into the evening sky, where it first appears as a crescent. Its phase slowly waxes as its sunlit hemisphere turns more in our direction. The Moon appears half-lit at First Quarter phase, so-called because it has completed a quarter of its orbit around Earth. The Moon continues waxing through its *gibbous* phase, until it appears totally lit at Full Moon and remains visible from sunset to sunrise.

The Moon's phases then play out in reverse. It wanes through its gibbous phase until it reaches Last Quarter, when it once again appears half-lit. This time, however, it's the opposite side that lies in sunlight, and the Moon appears strictly in the morning sky. The Moon steadily shrinks as a waning crescent until it disappears once again at New Moon.

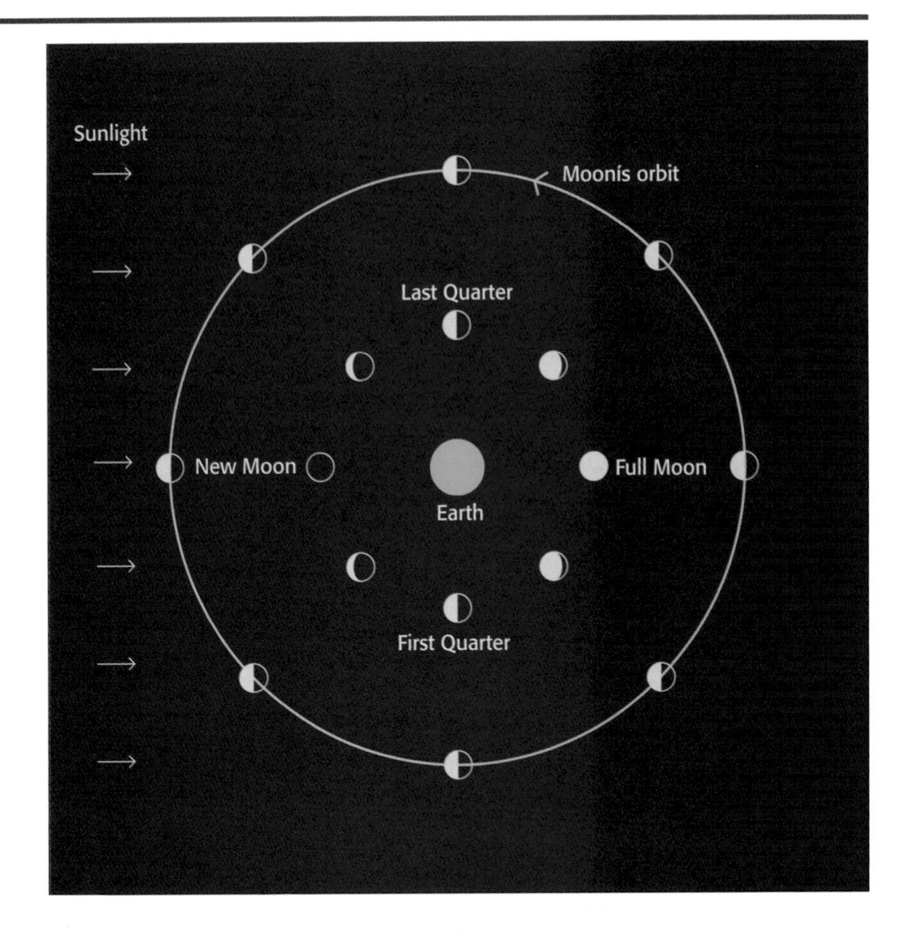

FAQ

Why do we see only one side of the Moon?

We see only one side of the Moon because it rotates on its axis in the same amount of time that it takes to revolve around Earth.

When the Moon first appears in the evening sky as a crescent, look carefully and you should see that its unlit part glows with a ghostly radiance. This is *earthshine,* or, more poetically, "the old Moon in the New Moon's arms." As the name implies, earthshine arises from sunlight that reflects off Earth's land, sea, and clouds, reaches the Moon, and then reflects back to Earth. When we see a crescent Moon, the Moon "sees" a nearly full Earth, and so a lot of our planet's light reaches the Moon. In this photo, we see earthshine on a waning crescent Moon.

CONTINUED ON NEXT PAGE

The Crescent Moon

When the Moon first appears in the western sky after sunset, its delicate crescent beckons observers. If you look through binoculars or a telescope at a 3- or 4-day-old Moon, the first feature you'll see is the dark expanse known as Mare Crisium. Closer inspection will reveal several intriguing craters, including Langrenus and Petavius. As always, detail appears sharpest along the terminator, where the Sun is rising and high ground casts long shadows.

Although not much of the Moon is lit at crescent phase, the circular "sea" known as Mare Crisium stands out. Early observers thought these dark expanses of lava were actually seabeds, and named them with the Latin word for sea: *mare* (pronounced MAH-ray; the plural is *maria* [MAH-ree-uh]). Mare Crisium (Sea of Crises) measures 350 miles across and formed when an asteroid smashed into the Moon's surface 3.85 billion years ago. Lava later flowed out to fill the basin.

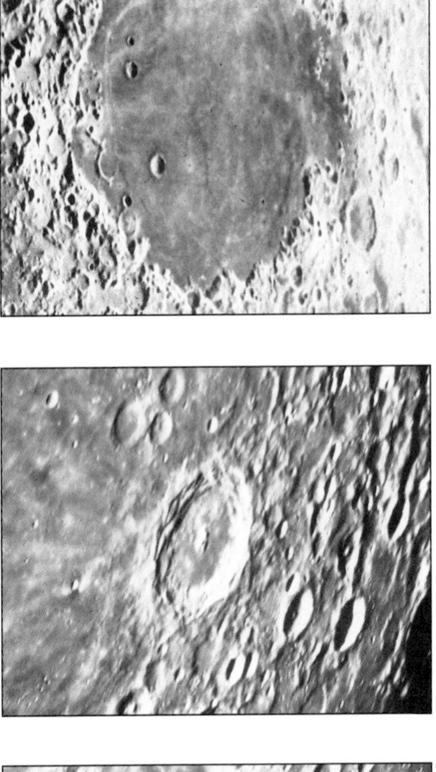

South of Mare Crisium lies Mare Fecunditatis (Sea of Fertility), and on its eastern shore lies the prominent crater Langrenus. This 80-mile-wide scar formed like almost all other lunar craters, when a meteorite slammed into the lunar surface. Such impacts often produce a central mountain peak when the crater floor rebounds. In Langrenus, a telescope will show you at least two central peaks. This crater also sports nicely terraced walls.

At the southeastern corner of Mare Fecunditatis, you'll find a 110-mile-wide crater named Petavius. This gouge features a chain of mountains at its center and a conspicuous *rille* (or groove) that runs from these mountains to the crater's rim. Such rilles developed after lava flooded the crater, then cooled and contracted. You can see other, fainter rilles on Petavius's floor during moments of good seeing just after the Sun rises there.

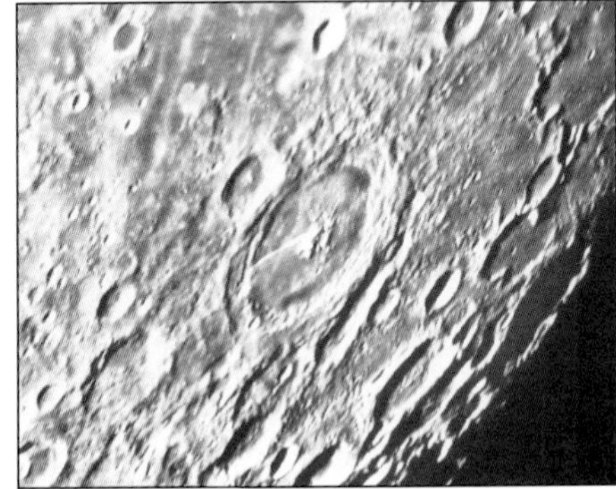

You never have a long time to view a crescent Moon. Once twilight fades enough for the Moon to stand out, it already lies fairly low in the sky. It also sets well before midnight, and so you have at most a few hours of good observing.

Note: *In all of these Moon images, north appears at top and lunar east to the right. This is the view binoculars will show to Northern Hemisphere observers when the Moon lies fairly high. A telescope likely will display a different orientation, but finding your way around a bright object like the Moon still proves fairly easy.*

TIP

Keep the Magnification Low

Turbulence in Earth's atmosphere rarely lets you use your telescope's highest magnification—particularly when you're viewing the Moon low in the sky. Start at low power, and then increase magnification until the image begins to waver.

CONTINUED ON NEXT PAGE

First Quarter Moon

The Moon's eastern hemisphere is on full display at First Quarter Moon. Several more maria come into view, including the oldest, Mare Nectaris (Sea of Nectar), at 3.92 billion years of age. At first quarter phase, we highlight three different types of features: the large gash of the Alpine Valley, the complex rilles near Triesnecker, and a trio of craters of noticeably different ages.

You don't have to travel to Europe to visit the Alps. Choose any clear night between First Quarter and Last Quarter Moon, and the lunar Alps (Montes Alpes) are in view. This mountain range forms the northeastern edge of Mare Imbrium (Sea of Rains). Cutting through the mountains is the 110-mile-long and 5-mile-wide Vallis Alpes (Alpine Valley). Take some time to explore this valley at different phases—you won't see a similar feature anywhere else on the Moon.

Three craters on the western shore of Mare Nectaris clearly show their relative ages. Theophilus, Cyrillus, and Catharina (from north to south) all span about 60 miles. Theophilus is the youngest and most spectacular. Its sharply defined rim and prominent terraces testify to its youth. Older Cyrillus appears more muted, and you can see where Theophilus destroyed part of its rim. The battered walls and floor of Catharina confirm it as the trio's most ancient.

The night of a First Quarter Moon is prime time for viewing the richest system of rilles on the Moon's surface. Just east of the 15-mile-wide crater Triesnecker lies a tangled web of a dozen or so grooves that formed as the lava in this region cooled and contracted. As a bonus, two other prominent rilles lie a little farther east. Rima Hyginus and Rima Ariadaeus both stretch some 135 miles and appear wider than the Triesnecker system.

The First Quarter Moon appears high in the south at sunset, and sets around midnight. It features the sites of the first and last human landings on the Moon. The first crew landed in southern Mare Tranquillitatis (Sea of Tranquility), and the last in southeastern Mare Serenitatis (Sea of Serenity). Don't even try to look for the American flags left behind—they and the lunar landers are far too small to see from Earth. Even the Hubble Space Telescope can't show them.

TIP

Out for a Jog

Around First Quarter Moon, the terminator seems to move at a rapid clip from night to night. But if you were on the lunar equator, you could keep pace with this line of sunrise at a good jogging speed of 10 mph.

CONTINUED ON NEXT PAGE

View the Lunar Surface *(continued)*

The Gibbous Moon

The waxing gibbous Moon gives us a first look at Mare Imbrium, an impact basin that spans 775 miles. Debris from this impact, which occurred 3.85 billion years ago, covers much of the Moon's nearside. Hundreds of millions of years later, lava oozed from below and filled the basin. At the southern end of Imbrium sits Copernicus, arguably the Moon's finest crater. Farther south, the rugged crater Clavius and the impressive Straight Wall vie for attention.

One of the largest craters on the Moon's nearside lurks deep in the heavily cratered southern highlands. Clavius spans 140 miles and appears better preserved than any other comparably sized crater. Clavius is so big that its floor sports several large craters. Five of these form a prominent arc, which starts with 30-mile-wide Rutherfurd and proceeds north and west to progressively smaller craters. Several lesser craters also pock Clavius's floor.

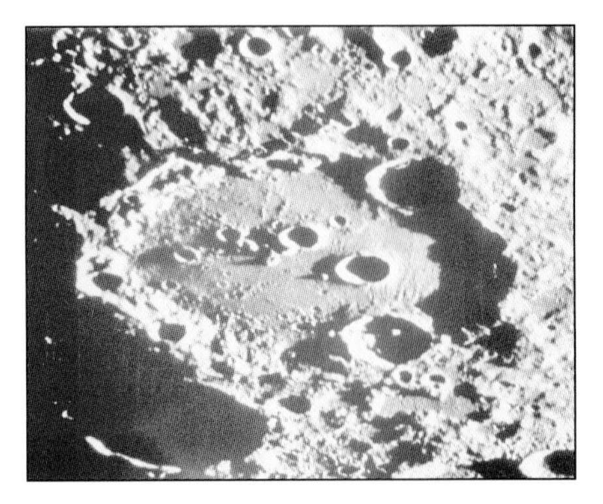

Rupes Recta, better known as the Straight Wall, sits on the eastern edge of Mare Nubium (Sea of Clouds). It first appears as a black streak a day or so after First Quarter Moon. The 70-mile-long "wall" rises only several hundred feet above the surrounding lava plains. Still, at sunrise, its black shadow appears conspicuous (left image). The Straight Wall disappears as Full Moon approaches, only to reappear as a white line a couple of days before Last Quarter (right).

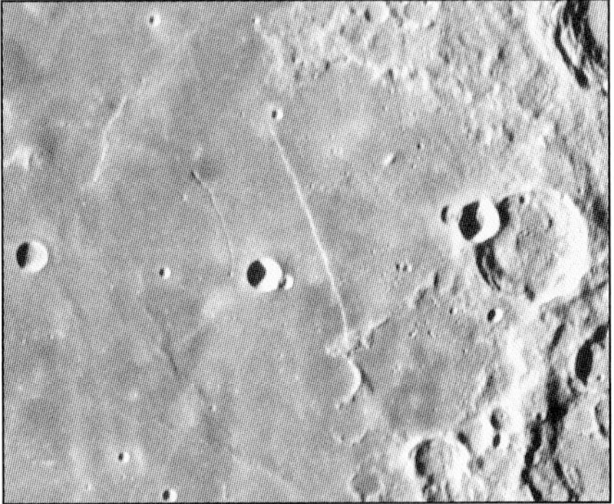

Magnificent Copernicus shows all the features of a fresh lunar crater. Formed in a giant impact some 800 million years ago (that's recent on the Moon), Copernicus measures 58 miles across. Huge landslides created the terraces that make up the crater's inner walls. Just after the Sun rises over the region, look for hundreds of secondary craters that surround Copernicus. These tiny craters formed when debris from the main impact crashed into the lunar surface.

A waxing gibbous Moon will light up the night from sunset until well past midnight. That's why deep-sky observers curse the Moon as it approaches full phase. It hurts the contrast and visibility of many deep-sky objects as effectively as light pollution does—and you can't drive away from the Moon. During the waxing gibbous phase, details on the Moon's eastern hemisphere become more muted because the Sun shines down from nearly overhead. The best views remain along the terminator, where night is turning to day.

CONTINUED ON NEXT PAGE

Full Moon

Although many people expect the Full Moon to provide the best viewing, the opposite is largely true. Because the Sun beats down from directly above, shadows and fine detail disappear. Yet the Full Moon excels at showing bright rays emanating from young craters like Tycho and Copernicus. Full Moon is also the time to view the Moon's brightest feature: the crater Aristarchus. Yet darkness also reigns at this phase, in the western basin Grimaldi.

The crater Tycho doesn't stand out for its size; it measures barely 50 miles across. And it doesn't show breathtaking detail at sunrise or sunset. But, at Full Moon, it's the most prominent feature on the Moon. A series of bright rays spread across the nearside and even wrap onto the farside. These rays are debris ejected by Tycho's relatively recent impact—only 108 million years ago. The dust hasn't had time to darken under the onslaught of solar radiation.

The brightest spot on the Moon is a crater only 25 miles in diameter. Aristarchus formed in a recent impact, which ejected pulverized rock that still shines brightly. It's so bright, in fact, that you can see it lit by earthshine a couple of days after New Moon. A large collapsed lava channel, called Vallis Schröteri (Schröter's Valley), starts at the crater Herodotus just west of Aristarchus. This is the easiest such feature to see through a small telescope.

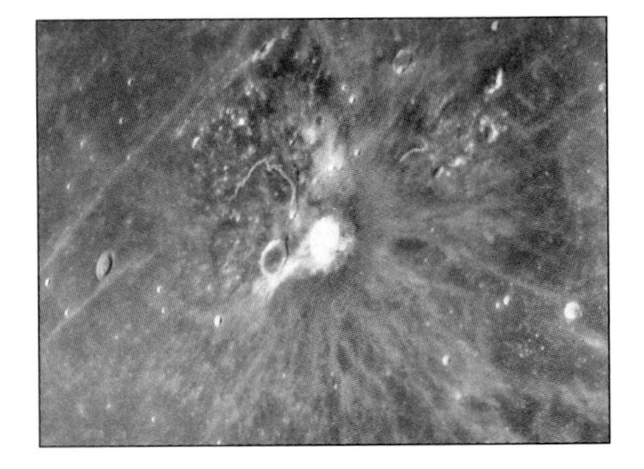

The Full Moon seems defined by brightness. It shines brilliantly all night long, and its most distinctive features are the bright rays emanating from relatively young impact craters. But lurking near the Moon's western limb, and visible from full phase until nearly New Moon, lies a particularly dark basin named Grimaldi. The lava inside Grimaldi appears darker than almost anywhere else on the Moon—and for reasons astronomers don't yet understand.

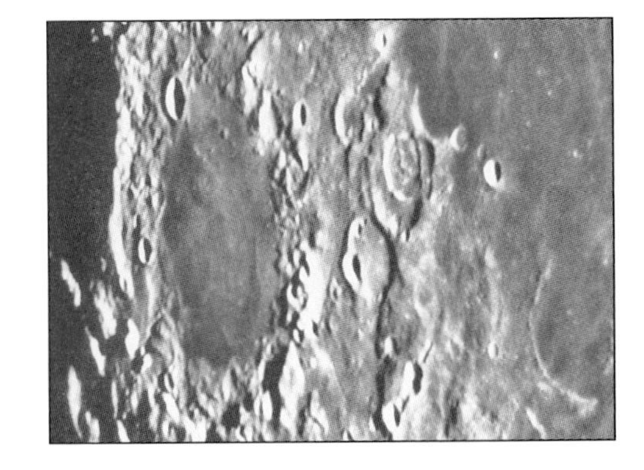

The Full Moon lies opposite the Sun in our sky, so it rises at sunset and sets at sunrise. Its glare fills the night sky and reduces deep-sky observers to viewing bright star clusters—and waiting for the Moon's phase to wane considerably. Lunar observers concentrate on the bright streaks of pulverized dust emanating from relatively fresh craters. The most spectacular system radiates from Tycho and extends more than halfway across the Moon's face.

TIP

Blinded by the Light

The Full Moon shines so brightly that viewing it through a modest-sized telescope can be uncomfortable. Try observing at higher power, which cuts down on the amount of light streaming through the eyepiece. Or simply don some sunglasses to reduce the glare.

Other Lunar Tricks

Although the Moon experiences the same phases each month, it doesn't always follow the same track across our sky. On spring evenings, the crescent Moon leaps higher from one night to the next. Then, on autumn evenings, the Full Moon lingers near the horizon for days on end. The Moon even tricks our perception. When a Full Moon lies near the horizon, it may appear much bigger than when it's higher in the sky—but it's really the same size.

For Northern Hemisphere observers, the crescent Moon appears highest in the sky on spring evenings. At this time of year, the *ecliptic*—the apparent path of the Sun across the sky that the Moon and planets closely follow—makes a steep angle to the western horizon after sunset. So, from one night to the next, the Moon's motion through the sky translates almost entirely into increased altitude. This makes spring the best time to hunt for a Moon a day or two old.

Move ahead to early autumn, and the ecliptic makes a shallow angle to the eastern horizon after sunset. From one night to the next, the Moon moves only a little farther below the horizon. And, instead of rising some 50 minutes later the next night as it typically does, the nearly Full Moon rises just 30 minutes later. The added light from the Moon used to help farmers bring in their crops, earning the early autumn Full Moon the name "Harvest Moon."

If you've ever watched a Full Moon rise, you'd probably swear it looked bigger than when it was higher. But the Moon is actually a little smaller near the horizon because we see it from a slightly greater distance. Scientists have yet to figure out this powerful illusion. Many believe our brains trick us because we see a low-hanging Moon near more familiar objects. To make the illusion disappear, face away from the Moon, bend over, and view it through your legs.

Eclipses of the Moon

When the Sun, Earth, and Moon line up exactly, and Earth lies in the middle, our planet's shadow falls on the Moon. The type of lunar eclipse depends on how precise the alignment is. If the Moon passes through only the outer part of the shadow, it's a partial eclipse. The eclipse becomes total if the whole Moon gets swallowed by the shadow. In that case, the Moon typically takes on a reddish cast—the light from all of Earth's sunrises and sunsets.

A lunar eclipse can happen only at Full Moon, when Earth lies between the Sun and the Moon. Most of the time, however, the Full Moon passes north or south of Earth's shadow. Usually every sixth Full Moon brings an eclipse, which can be either total or partial depending on how close the Moon travels to the shadow's center. The good news: Anyone with a clear sky on Earth's night side can see a lunar eclipse, so about half the world gets in on the action.

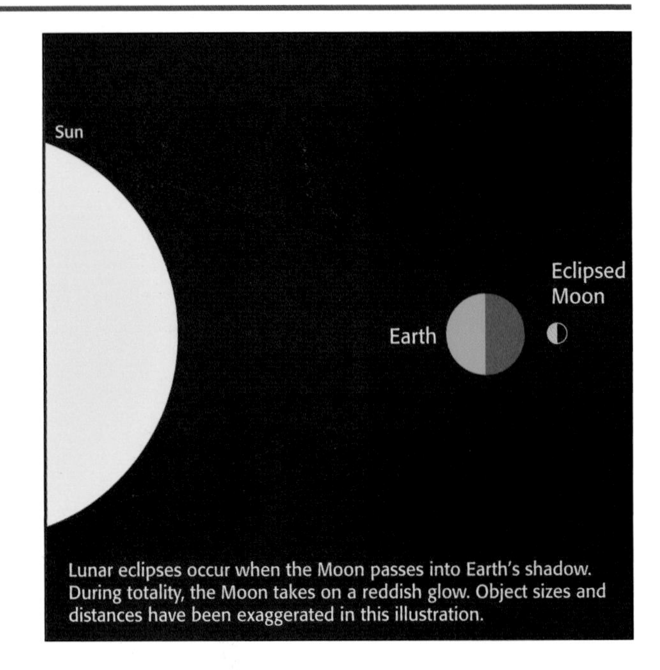

Lunar eclipses occur when the Moon passes into Earth's shadow. During totality, the Moon takes on a reddish glow. Object sizes and distances have been exaggerated in this illustration.

Although partial eclipses lack the grandeur of total ones, they can still be exciting to watch. Lunar eclipses usually last for a few hours from start to finish, so you have time to sit back and enjoy the view.

Watching a total lunar eclipse can leave you spellbound. To capture the feeling, try taking a wide-angle shot of the scene. The best photos capture a picturesque foreground with the eclipsed Moon as a backdrop. In this image from February 2008, the eclipsed Moon stands high above the Temple of Poseidon in Greece.

CONTINUED ON NEXT PAGE

During the partial phases of a lunar eclipse, Earth's shadow looks pretty dark against the brightly lit lunar surface. Even though some light penetrates our planet's shadow, the stark contrast makes it hard to see much on the eclipsed part of the Moon.

As totality approaches, the part of the Moon's surface in shadow starts to take on a reddish hue, while the still-sunlit lunar surface gleams brightly. Although lunar eclipses progress slowly, the Moon's appearance changes throughout. Even if cold temperatures or biting mosquitoes make you seek shelter, it's worth checking on the eclipse's evolution every five or ten minutes.

A sequence of images from the August 2007 total lunar eclipse shows that Earth's shadow spans more than twice the Moon's diameter. If the Moon passes near the shadow's center, you get a long eclipse. The total phase for this eclipse lasted 90 minutes, while the entire eclipse took three-and-a-half hours to play out. The longest total phases of lunar eclipses last more than 100 minutes.

A close-up view of the Moon during a total eclipse shows it bathed in reddish light. The glow comes from sunlight that has passed through Earth's atmosphere. The air scatters out blue light and leaves only red—the same reason sunrises and sunsets look red—and bends this remaining light into the shadow. Depending on the state of Earth's atmosphere, the eclipsed Moon's color can range from bright orange to deep red, and occasionally the Moon nearly disappears.

Note: *In 2015 (see table at right), the total lunar eclipse starts the evening of the 27th and ends the morning of the 28th.*

Total Lunar Eclipses through 2017

Date	*Duration*
December 21, 2010	73 minutes
June 15, 2011	101 minutes
December 10, 2011	52 minutes
April 15, 2014	79 minutes
October 8, 2014	60 minutes
April 4, 2015	12 minutes
September 27/28, 2015	73 minutes

Eclipses of the Sun

When the Sun, Moon, and Earth line up exactly, with the Moon in the middle, the Moon's shadow falls on Earth. The type of solar eclipse depends on the alignment's precision and on the Moon's distance from Earth. If Earth glides through the outer part of the Moon's shadow, we see a *partial* eclipse. A *total* eclipse happens when the Moon's dark shadow touches Earth, while an *annular* eclipse results if the Moon is too far away to block the whole Sun.

A solar eclipse can happen only at New Moon, when the Moon lies between the Sun and Earth. At most New Moons, however, Earth passes north or south of the shadow. The darkest part of the Moon's shadow, called the *umbra,* is a narrow cone that doesn't always reach Earth. The much wider, and lighter, penumbral shadow can cover a good portion of Earth's daytime hemisphere.

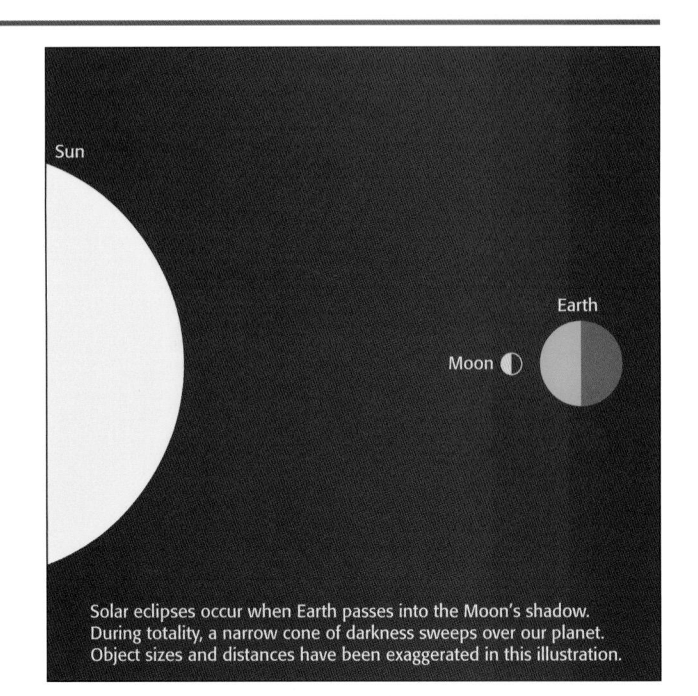

Solar eclipses occur when Earth passes into the Moon's shadow. During totality, a narrow cone of darkness sweeps over our planet. Object sizes and distances have been exaggerated in this illustration.

If the Moon lies relatively close to Earth during a central eclipse, its umbral shadow will reach our planet's surface. The dark shadow can be anywhere from a pinpoint to an oval more than 100 miles across. The wider the shadow, the longer totality can be. If the Moon lies too far away during a central eclipse, it can't block the entire Sun, and we see an annular eclipse instead. If Earth passes only through the penumbra, we get a partial solar eclipse.

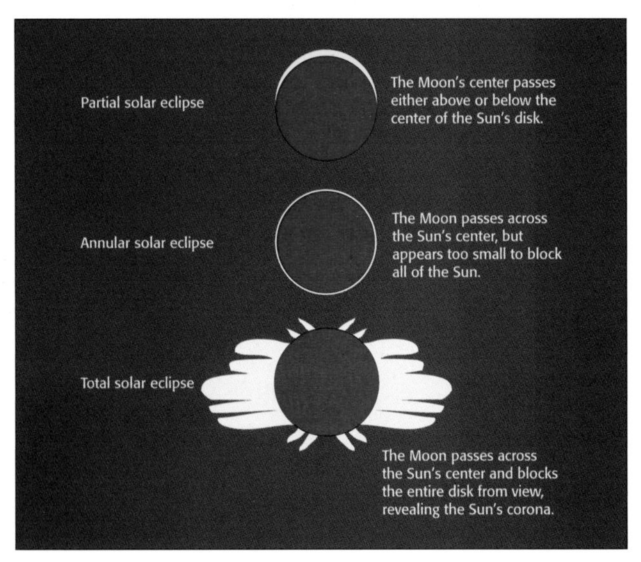

Seeing a total solar eclipse will take your breath away—and likely leave you thinking about when you can see another one. The path of totality often traverses exotic locales, offering dramatic scenery for eclipse photography. The March 2006 eclipse passed through central Turkey, where a cluster of "fairy chimneys" (volcanic rock formations) provided a foreground for the totally eclipsed Sun.

TIP

Watch Your Eyes

Viewing a solar eclipse risks your eyesight just as much as observing the Sun does. You must properly protect your eyes throughout the eclipse's partial phases. Once totality arrives, however, and the Moon blocks the Sun's bright surface from view, you can watch the eclipse without a filter.

CONTINUED ON NEXT PAGE

Eclipses of the Sun *(continued)*

A partial solar eclipse looks nice, but it won't excite people anywhere near as much as a total eclipse. The partial phases of a total eclipse, however, are another story. In this case, anticipation builds as the Moon covers more of the Sun's disk. Watch as shadows grow sharper, the temperature begins to drop, and colors in the landscape become more saturated. And by the way, even if you've seen a 99-percent partial eclipse, it doesn't come close to totality.

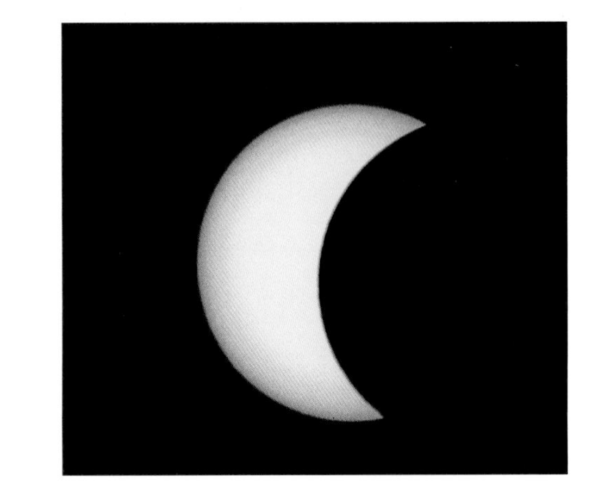

Just as a pinhole camera projects an image of the Sun, the tiny gaps in the leaves of a tree or bush become pinholes to project solar images. During an eclipse's partial phases, you can see hundreds or thousands of crescents projected on the ground as sunlight filters through the leaves. It's a sight you'll want to look for, although if you're anxiously awaiting totality, wait until after the main event.

An annular solar eclipse occurs when the Moon passes directly between the Sun and Earth but lies too far away to cover the whole Sun. You're left with a thin ring of sunlight surrounding the Moon's black disk. (The word "annular" comes from the Latin *annulus*, meaning "ring.") Annular eclipses can be memorable events, but they still don't come close to a total solar eclipse.

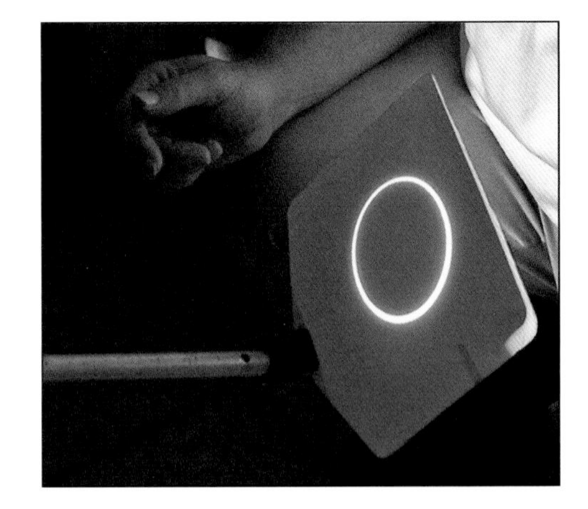

The reason people travel thousands of miles to see totality is to get a fleeting glimpse of the Sun's corona. The corona's gauzy, pearly-white glow typically stretches at least twice the Moon's diameter. Loops, swirls, and streamers often punctuate the corona, which shines roughly half as bright as a Full Moon. It's normally invisible simply because the Sun shines so brightly. The rarefied corona is the outermost and hottest part of the solar atmosphere.

CONTINUED ON NEXT PAGE

Eclipses of the Sun *(continued)*

Just before totality arrives, mountains along the Moon's limb start to poke through the Sun's dwindling crescent. Soon, sunlight only passes through deep valleys along the Moon's advancing limb. This string of bright points, known as Baily's beads, heralds the imminent approach of totality.

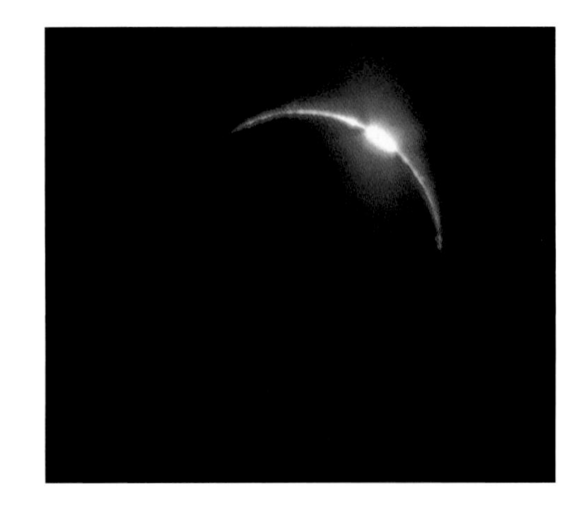

The longest total solar eclipse of the next decade occurs July 22, 2009. Eclipse chasers will converge on eastern Asia and the western Pacific for the best views. Totality lasts more than five minutes in Shanghai. At maximum eclipse in the Pacific, observers will see more than 6 minutes and 30 seconds of totality.

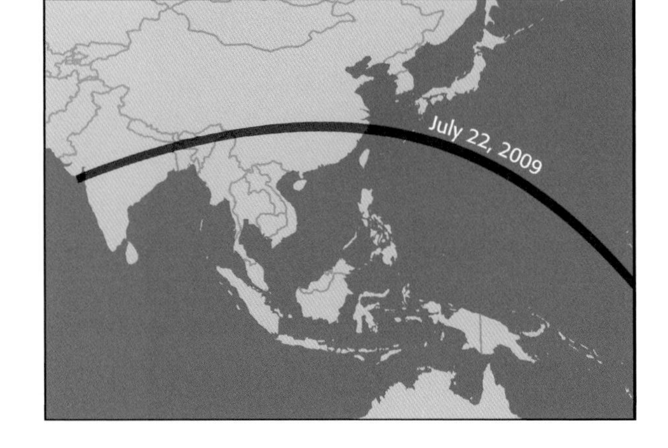

Eclipse fans in North America have had August 21, 2017, circled on their calendars for ages. On that date, for the first time since February 1979, the Moon's umbral shadow will touch the continental United States. The path of totality stretches from Oregon to South Carolina, and anyone on the center line will experience at least two minutes of totality.

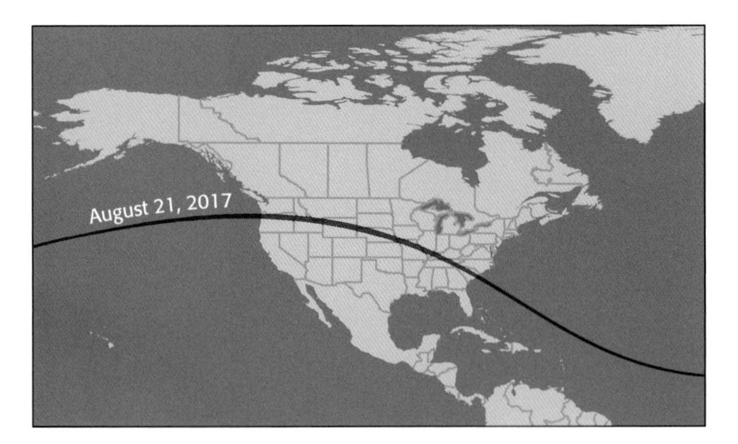

As totality approaches, the jewel-like Baily's beads dwindle to a lone solitaire: the aptly named diamond ring. Once the ring disappears, totality has started and the solar filters can come off. Then it's time to explore the corona and the fiery red prominences arcing above the limb. Totality lasts at most a few minutes, but these are minutes you'll never forget.

Total Solar Eclipses through 2017

Date	*Maximum Duration*	*Location*
July 22, 2009	6 minutes and 39 seconds	eastern Asia
July 11, 2010	5 minutes and 20 seconds	South America
November 13, 2012	4 minutes and 2 seconds	Australia
November 3, 2013	1 minute and 40 seconds	Africa
March 20, 2015	2 minutes and 47 seconds	North Atlantic
March 9, 2016	4 minutes and 9 seconds	Indonesia
August 21, 2017	2 minutes and 40 seconds	North America

chapter 9

Observe the Rest of the Solar System

Viewing the planets presents different challenges and rewards. First, for the most part, the planets shine brightly. This makes them easy to find. But the planets also move across the sky relative to the stars, so you can't locate them in one place all the time. (The word *planet* comes from a Greek word meaning "to wander.")

Seeing a planet through a telescope brings a thrill to most backyard observers. These are real worlds visited by spacecraft, and most people have heard or read about them since their school days. And Saturn, of course, looks simply incredible. Many people swear they're seeing a photograph when they first view Saturn up-close.

The Inner Planets: Mercury and Venus

For thousands of years, skywatchers recognized five points of light that behaved differently from all the rest. These so-called planets moved relative to the background stars. Two of them—Mercury and Venus—appeared only in the west after sunset (as "evening stars") or in the east before sunrise (as "morning stars"). Mercury never lies far from the Sun and is tricky to locate. Venus shines brilliantly and stands out most of the time.

Because Mercury and Venus orbit closer to the Sun than Earth, they can never appear far from the Sun's glare in our sky. The best time to observe either planet is when it lies farthest from our star, at greatest eastern or western elongation. You can't view the inner planets during the brief period when they pass between the Sun and Earth *(inferior conjunction)* or the longer period when they lie on the Sun's far side *(superior conjunction).*

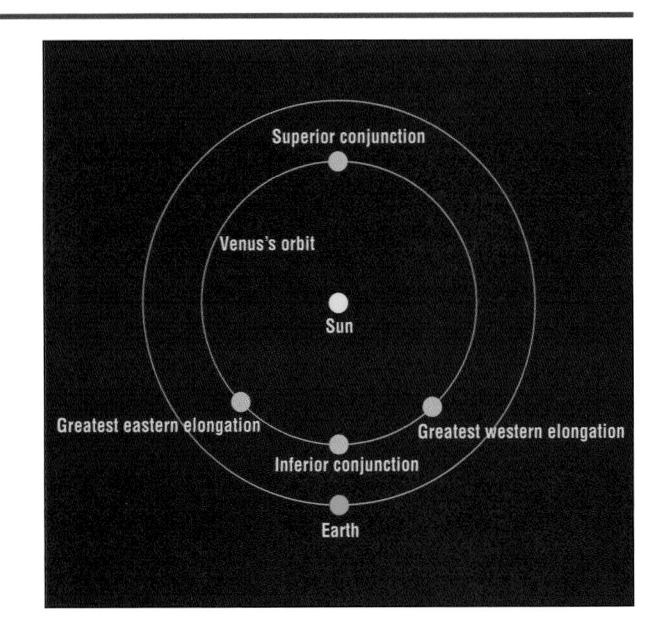

The best times to view an inner planet through a telescope come from greatest eastern elongation to shortly before inferior conjunction, and again from shortly after inferior conjunction to greatest western elongation. During both of these periods, the planet looms larger and its crescent phase changes rapidly. (Both planets appear half-lit around greatest elongation.)

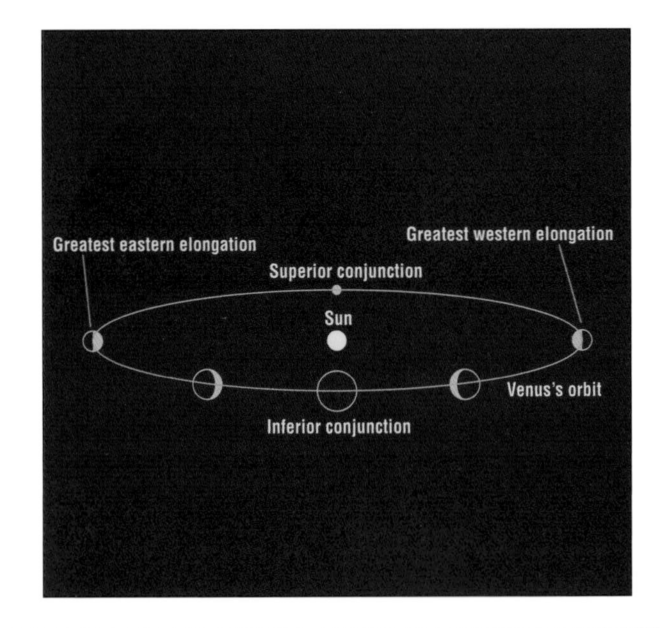

Mercury

Mercury reveals few details to observers on Earth. Its small size and proximity to the Sun keep us from getting a clear look. When viewing Mercury through a telescope, be content with watching its changing size and phase. This view shows the innermost planet on March 3, 2008, when it appeared 7 arcseconds in diameter and barely more than half-lit.

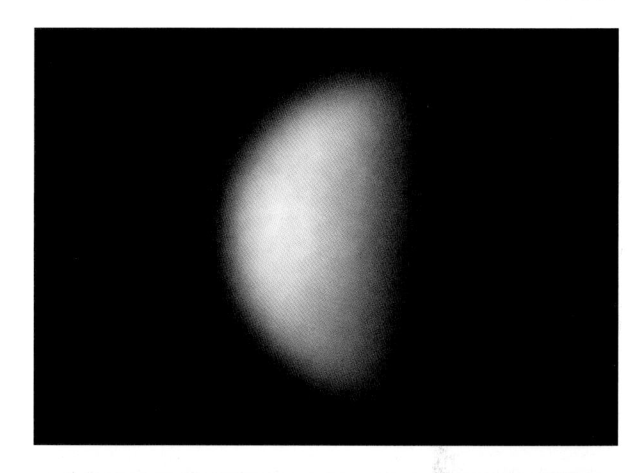

To see Mercury's surface, you need to send a spacecraft to it. NASA's MESSENGER probe snapped this image in January 2008, when it came within 125 miles of the planet. It shows a cratered landscape similar in many ways to the Moon, but with intriguing differences. MESSENGER will fly past Mercury twice more before it enters orbit around the planet in 2011.

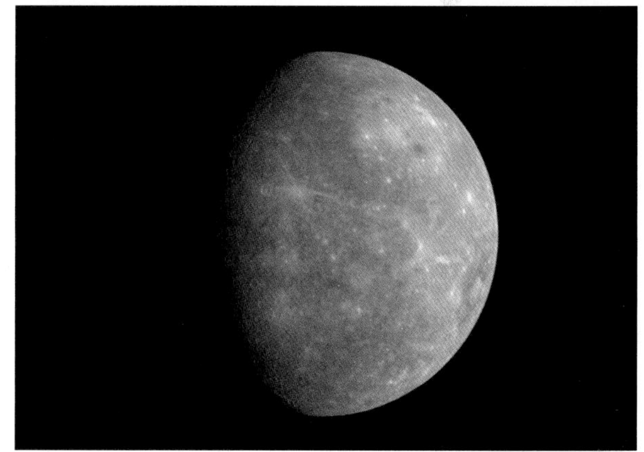

Mercury's Finest Elongations for Northern Observers

Date of Greatest Elongation	*Morning or Evening Sky*	*Date of Greatest Elongation*	*Morning or Evening Sky*
April 26, 2009	Evening	August 16, 2012	Morning
October 5, 2009	Morning	June 12, 2013	Evening
April 8, 2010	Evening	November 17, 2013	Morning
September 19, 2010	Morning	May 25, 2014	Evening
March 22, 2011	Evening	November 1, 2014	Morning
December 22, 2011	Morning	May 7, 2015	Evening
March 5, 2012	Evening	October 15, 2015	Morning

CONTINUED ON NEXT PAGE

Venus

Venus's size and appearance from Earth changes rapidly as it moves from inferior conjunction toward greatest western elongation. This telescopic image shows the planet on October 20, 2007, one week before greatest elongation. Venus appears 26 arcseconds across and sports a distinct crescent phase. (When Venus becomes visible after inferior conjunction, it can appear twice as large.)

At greatest western elongation on October 28, 2007, Venus appeared half-lit and 24 arcseconds across—nearly 10 percent smaller than it did just a week previously (above). In the months following greatest western elongation, Venus's apparent size continues to shrink while its phase waxes, but the changes proceed at a more leisurely pace.

On rare occasions, you can glimpse Mercury and Venus at inferior conjunction. If the alignment between the Sun, Earth, and planet is precise, the planet will transit the Sun's disk.

Note: *To view the planet's silhouette, be sure to use proper eye protection (see Chapter 8).*

This photo shows the dark disk of Venus transiting the Sun on June 8, 2004. The next Venus transit—and the final one this century—occurs June 6, 2012.

Unfortunately for earthbound observers, thick clouds shroud Venus's surface. The only details a telescope will reveal are the planet's changing size and phase. To see the surface, scientists probe Venus with radar. This view from the Magellan spacecraft displays one hemisphere of Venus. The planet shows raised "continents" rising above lower surroundings. With a surface temperature hot enough to melt lead, no water fills these basins.

The large volcano Maat Mons rises some 3 miles above Venus's surrounding plains, while lava flows extend more than 100 miles. Scientists created this perspective view from Magellan radar observations. The landscape colors are based on observations made from Venus's surface by Soviet Venera spacecraft.

Upcoming Greatest Elongations of Venus

Date of Greatest Elongation	*Morning or Evening Sky*	*Date of Greatest Elongation*	*Morning or Evening Sky*
January 14, 2009	Evening	August 15, 2012	Morning
June 5, 2009	Morning	November 1, 2013	Evening
August 19, 2010	Evening	March 22, 2014	Morning
January 8, 2011	Morning	June 6, 2015	Evening
March 27, 2012	Evening	October 26, 2015	Morning

Mars: The Red Planet

No planet fascinates us more than Mars. The Red Planet shows details through earthbound telescopes, and those details change with time. Some early observers interpreted the changes as signs of vegetation. Although we now know plants don't grow on Mars, scientists have learned that this most Earthlike planet once had flowing water, and life perhaps could have gained a foothold. Even a small telescope will show white polar caps and dark surface markings.

Because Mars and the other outer planets orbit farther from the Sun than Earth, they can appear anywhere along the ecliptic relative to the Sun. The best time to view an outer planet is near the time of opposition, when it lies opposite the Sun in our sky and remains visible all night. Opposition also marks the time when a planet appears brightest and looms largest through a telescope's eyepiece.

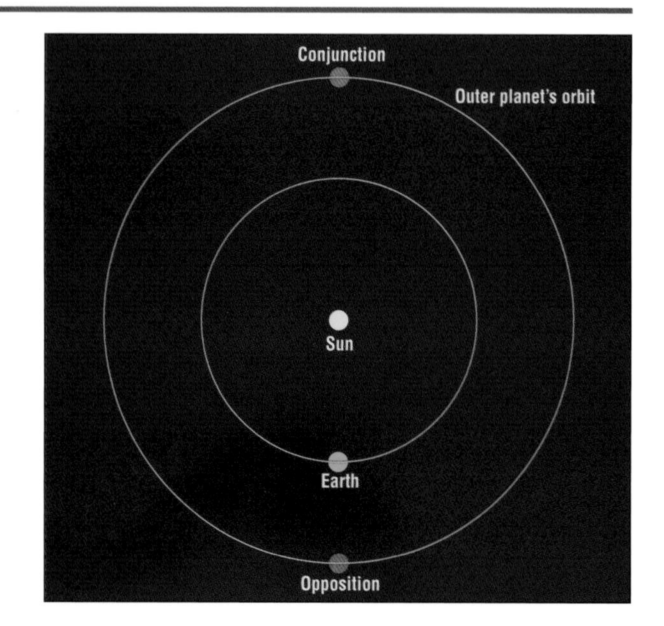

Mars has a more elongated orbit than any other outer planet. For observers, this means Mars appears much better at some oppositions than at others. If Mars lies close to the Sun at opposition, it will loom large through a telescope. If Mars reaches opposition on the outer part of its orbit, it appears small. The Red Planet's best recent opposition occurred in 2003. It will hit bottom in 2012, and then start growing larger again beginning with its 2014 opposition.

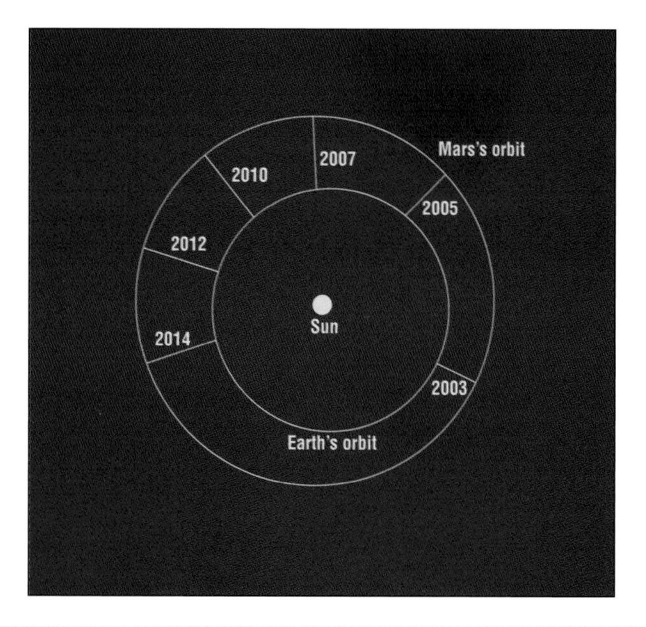

As the closest of the outer planets, Mars spends more time between oppositions than any other planet—slightly more than two years. So, the Red Planet takes several months to move from conjunction to opposition, growing bigger and brighter at a snail's pace. In this view from early June 2007, Mars appears only 6 arcseconds across and shows little detail beyond its south polar cap (at bottom).

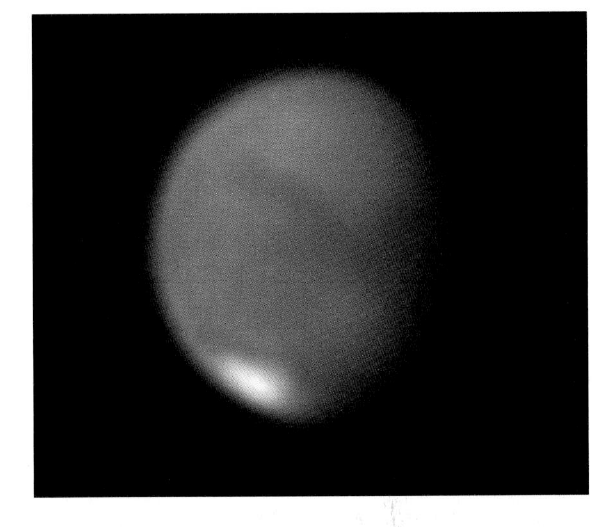

As Mars approached opposition in late 2007, it appeared much bigger than above. This photo shows the planet in late November 2007, when its diameter had ballooned to 15 arcseconds. Now its north polar cap (top) appears prominent as winter there nears its end. Also note how the dark features, which trace the planet's shifting sands, become conspicuous on the planet's larger disk.

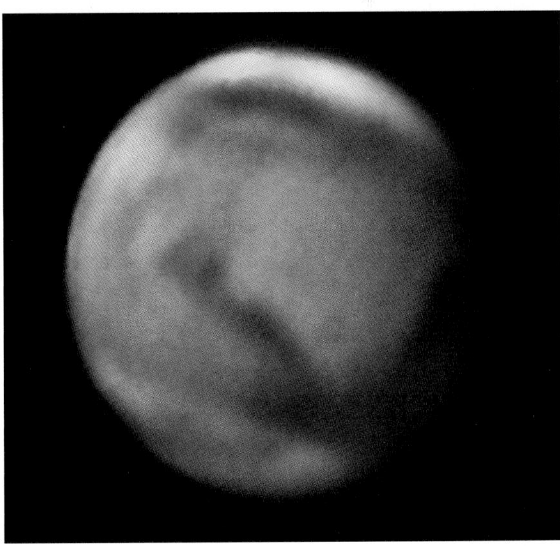

CONTINUED ON NEXT PAGE

TIP

Seeing Mars's Changing Face

The view of Mars through a telescope changes from hour to hour and day to day as the Red Planet rotates on its axis. Mars takes 24 hours and 37 minutes to complete a rotation, so you can see much of the surface in just one long night. If you observe at the same time each night, Mars's surface features will shift slowly eastward.

Mars: The Red Planet *(continued)*

The best views of Mars ever seen from Earth came at the August 2003 opposition. Mars was then just under 35 million miles away—its closest approach in nearly 60,000 years. Astronomers used the sharp eye of the Hubble Space Telescope to examine the Red Planet during this historic opposition. In this view, the prominent features include the bright south polar cap (at bottom), the circular Hellas impact basin to its upper right, and the dark Syrtis Major at top right.

Just 11 hours after it took the Mars portrait above, the Hubble Space Telescope captured the planet's opposite hemisphere. The bright oval seen above the center of this view is Olympus Mons, the solar system's largest volcano. This mammoth mountain spans 360 miles and rises 9 miles above its surroundings. The orange streaks seen on the south polar cap mark where Martian dust has blown across the icy expanse.

Although spacecraft orbiting Mars have returned stunning images of the surface, nothing quite compares with the view from ground level. NASA's Opportunity rover, which landed on Mars in January 2004, took this true-color image of Endurance Crater and the windswept sand dunes in its bowl (see below). Scientists long suspected Mars had water on its surface in the distant past, but Opportunity and its twin, Spirit, proved the case.

Upcoming Mars Oppositions		
Date of Opposition	***Peak Magnitude***	***Constellation***
January 29, 2010	–1.3	Cancer
March 3, 2012	–1.2	Leo
April 8, 2014	–1.5	Virgo

Jupiter: King of the Planets

It takes special qualities to be compared with royalty, and Jupiter delivers. More than a thousand Earths would fit inside Jupiter, and the giant planet contains twice as much material as the other planets combined. Look at Jupiter through a telescope, and all you'll see are clouds. But unlike Venus's bland clouds, Jupiter's show tremendous detail. You'll see dark belts, light zones, and an array of other features. You'll also spy the planet's four bright moons.

Jupiter's cloud tops normally explode with activity. On most clear nights from Earth, a small telescope will show two dark equatorial belts sandwiching a brighter equatorial zone. On nights with little turbulence in our atmosphere, you can see a series of alternating belts and zones. Turbulence in Jupiter's atmosphere typically shows up along the edges of the dark belts. Look for loops, bumps, and projections, all of which can change appearance from night to night.

Jupiter's Great Red Spot—a vast storm bigger than Earth—has raged for more than three centuries. If it happens to be on the half of Jupiter facing Earth, a small telescope usually shows it. Recently, the Red Spot has had company. In 2006, Red Spot Jr. formed a bit south of its big brother. And in 2008, a third red spot appeared in the other two's vicinity. Scientists expect the Great Red Spot to swallow this newest neighbor within a couple of years.

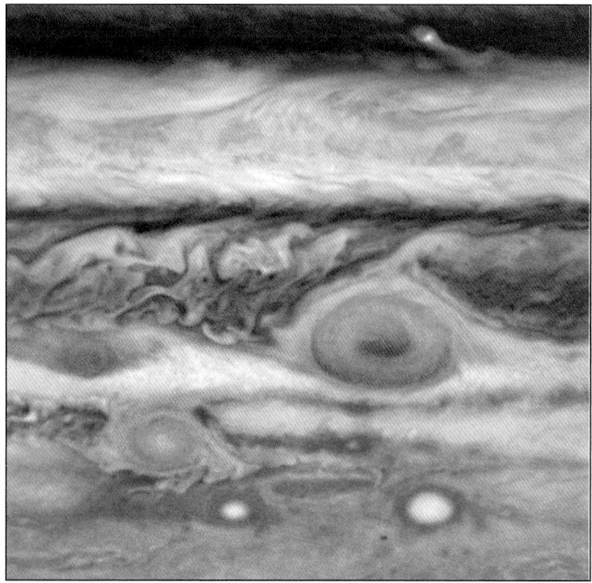

The broad patterns of belts and zones in Jupiter's atmosphere stay consistent from one month to the next, but smaller features can change shape daily. With a rotation period of less than 10 hours, Jupiter will reveal most or all of its cloud tops in a single night near opposition. The three photographs on this page, all taken in 2007 through a 12-inch telescope, show different faces of the planet. This view highlights the Great Red Spot.

In this view, the Great Red Spot appears near the planet's limb. It had just rotated onto Jupiter's visible hemisphere. About two hours after this image was taken, the spot appeared at the center of the disk, and two hours after that, it was starting to rotate out of view on the opposite side.

Our final view shows the hemisphere of Jupiter without the Red Spot. Note the deep brown hue of the North and South Equatorial Belts, and the turbulence that appears near their edges. In all three of these photos, Jupiter appeared more than 40 arcseconds across its equator. Look carefully, however, and you'll notice Jupiter's polar diameter is less. The giant planet bulges thanks to its gaseous nature and rapid rotation.

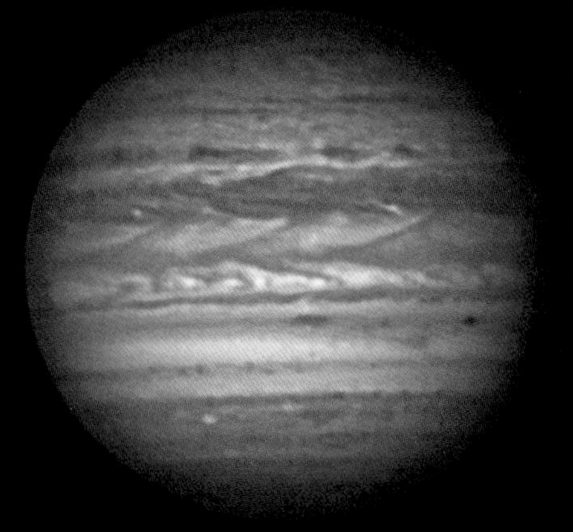

CONTINUED ON NEXT PAGE

Jupiter: King of the Planets *(continued)*

Jupiter's atmosphere presents an ever-changing tapestry, but its four bright moons put on an equally impressive show. Discovered by Galileo when he first turned his telescope toward Jupiter, these Galilean moons endlessly circle the planet. Their positions change radically from one night to the next. You can even see the motions of the faster inner moons over the course of a few hours. In this view, the dark shadow of Callisto falls on Jupiter's cloud tops.

The stark black shadows of the Galilean moons stand out against the bright clouds of Jupiter. In these views, both innermost Io and its shadow show up. In the earlier image (left), the shadow lies near the center of Jupiter's disk while Io itself lies directly on the planet's left-side limb. Less than an hour later (right), the shadow remains conspicuous but the orange-colored moon, now projected against the cloud tops, has nearly disappeared.

More often than not, you'll see all four moons separate from the planet—and often in a line. The moons appear as bright points of light. They all glow bright enough that they would be visible with the naked eye if not for Jupiter's glare. In this view from May 2007, innermost Io lies farthest to the left, with the biggest moon, Ganymede, between it and Jupiter. On the opposite side, outermost Callisto lies above the smallest Galilean moon, Europa.

This view from the Cassini spacecraft, taken as it flew past in late 2000 on its way toward Saturn, paints the most-detailed global portrait of Jupiter yet. The Great Red Spot stands out at the southern edge of the South Equatorial Belt, while fine features show up in all the other belts and zones. Although this true-color image shows more detail than we can see from Earth, ground-based telescopes can do a remarkably good job.

Upcoming Jupiter Oppositions

Date of Opposition	*Peak Magnitude*	*Constellation*
August 14, 2009	–2.9	Capricornus
September 21, 2010	–2.9	Pisces
October 28, 2011	–2.9	Aries
December 2, 2012	–2.8	Taurus
January 5, 2014	–2.7	Gemini
February 6, 2015	–2.6	Cancer

Saturn: Ringed Wonder

Saturn ranks at or near the top of every observer's list of favorites. One look through a telescope will tell you why: An image of the ringed planet looks like it was plucked from a picture book. First-time observers sometimes expect more from the sky than it delivers. Saturn invariably exceeds people's expectations. Although the ring system garners most of the attention, Saturn also has a family of moons worth exploring.

The first thing everyone notices when they look at Saturn is its magnificent system of rings. The rings span more than 150,000 miles yet measure less than a mile thick. One gap appears obvious from Earth: the Cassini Division separating the outer A ring from the brighter B ring. Under excellent observing conditions, a faint inner ring, called the C ring, also appears, as does the narrow Encke Division near the A ring's outer edge.

TIP

Where to Look for Planets

Although Venus shines so brightly it's hard to miss, some of the other planets can be difficult to recognize against the background stars. To find these interlopers, look at the large, circular star maps that open Chapters 4 through 7. On each map, a dashed arc represents the ecliptic—the path of the planets across the sky. Simply look along the ecliptic for a bright object that doesn't belong. Odds are, it's a planet.

Through the eyes of the Cassini spacecraft orbiting Saturn, the broad rings visible from Earth resolve into thousands of narrow ringlets. The rings are not solid, but consist of untold numbers of bodies ranging from tiny bits of rock and ice up to house-sized boulders. Notice that even with Cassini's close-up view, Saturn's atmosphere displays little detail. The same kind of turbulence visible in Jupiter's cloud tops likely exists on Saturn, but haze hides it.

Saturn possesses one of the solar system's most remarkable moons. Titan, seen here in a Cassini close-up (with Saturn in the background), has a nitrogen-rich atmosphere thicker than Earth's. Cassini's radar instrument found lakes of liquid methane on the planet-sized moon. This makes Titan the only object other than Earth known to harbor liquids on its surface. Titan makes an easy target for earthbound telescopes, and several other moons show up with little trouble.

CONTINUED ON NEXT PAGE

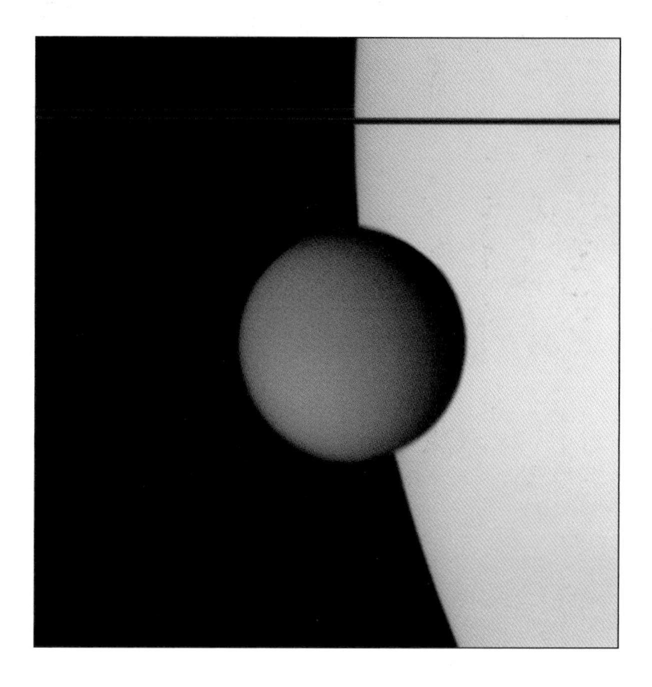

TIP

Spying Saturn's Moons

Unlike Jupiter's moons, which shine brightly and often form a straight line, Saturn's moons pose an observing challenge. Titan appears bright enough to see easily in any telescope, but you'll need a 6-inch telescope to reliably see Tethys, Dione, and Rhea. And the next tier of moons—Mimas, Enceladus, and Iapetus—may require 10 inches of aperture.

Saturn: Ringed Wonder *(continued)*

The photos on these two pages show Saturn's rings through the full range of tilt visible from Earth. When the rings are nearly edge-on, they appear as a thin strip with little detail. As the rings open, the Cassini Division and other features show more clearly. The rings will tilt edge-on in 2009 and be difficult or impossible to see. This view shows the rings in October 1996. We'll see a near repeat at the 2010 opposition.

This view shows the rings as they appeared in October 1997. We'll see a repeat performance at opposition in 2011.

This view shows the rings as they appeared in October 1998. We'll have a similar view at opposition in 2012.

This view shows the rings as they appeared in November 1999. The rings will have a similar tilt at opposition in both 2013 and 2014.

This view shows the rings as they appeared in November 2000—about as wide open as they ever get. We'll see a repeat performance at opposition in 2015.

Upcoming Saturn Oppositions

Date of Opposition	*Peak Magnitude*	*Constellation*	*Ring Tilt*
March 8, 2009	0.5	Leo	3°
March 21, 2010	0.5	Virgo	4°
April 3, 2011	0.4	Virgo	11°
April 15, 2012	0.2	Virgo	17°
April 28, 2013	0.1	Libra	22°
May 10, 2014	0.1	Libra	26°
May 22, 2015	0.0	Libra	29°

The Outer Planets: Uranus and Neptune

If you look at family traits, Uranus and Neptune would be the solar system's twins. Both are gas giants with roughly 15 times Earth's mass. They also sport a similar bluish color when seen through a telescope. Although astronomer William Herschel discovered Uranus telescopically in 1781, it glows bright enough to see with the naked eye from a dark site. More-distant Neptune, a 19th-century discovery, can't be seen without some sort of optical aid.

Uranus

If you have a dark sky and know where to look, you can spot Uranus without optical aid. Binoculars or a telescope pick it up easily. So why didn't anyone discover it before William Herschel? To the naked eye, Uranus looks like a faint star—nothing sets it apart from the ordinary. Herschel saw the planet's small disk and its motion relative to the background stars. Observers with telescopes today easily notice the planet's distinct blue-green color.

It takes a large telescope to see detail on Uranus. Here, the Hubble Space Telescope reveals the planet's banded atmosphere and some brighter clouds. The bands look similar to those seen in greater detail on Jupiter and Saturn. Hubble also shows the planet's complex ring system. Unlike Saturn's icy rings, which reflect a lot of sunlight, Uranus's dusty rings barely show up.

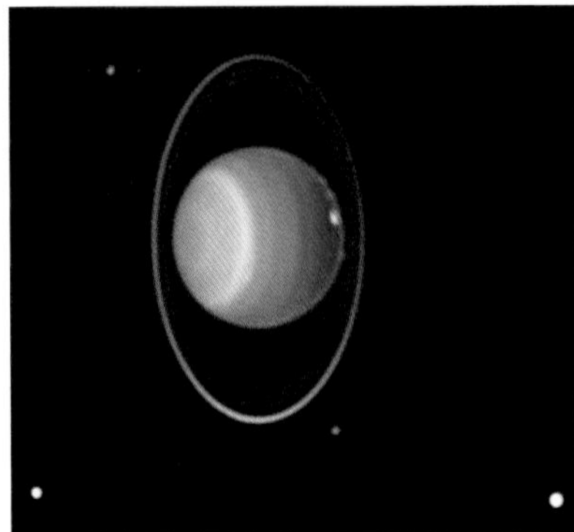

Upcoming Uranus Oppositions		
Date of Opposition	***Peak Magnitude***	***Constellation***
September 17, 2009	5.7	Pisces
September 21, 2010	5.7	Pisces
September 25, 2011	5.7	Pisces
September 29, 2012	5.7	Pisces
October 3, 2013	5.7	Pisces
October 7, 2014	5.7	Pisces
October 11, 2015	5.7	Pisces

Neptune

The gas-giant Neptune is the farthest planet from the Sun. Unlike its twin, Uranus, Neptune reflects too little sunlight for it to show up without optical aid. Still, binoculars and small telescopes bring it within range of backyard observers. Don't expect to see anything beyond a tiny disk and a subtle blue-gray color. The cloud patterns visible in this Hubble image show up only through professional telescopes.

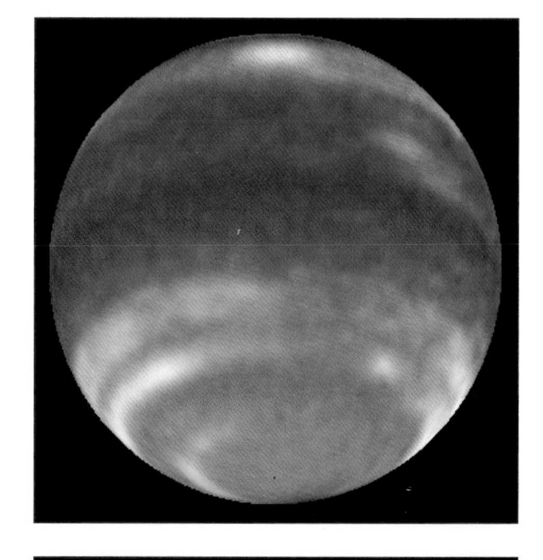

Humans have seen Neptune close-up only once, when the Voyager 2 spacecraft flew past in 1989. Voyager revealed a surprisingly storm-ravaged world, where winds occasionally gust up to 900 miles per hour. Although the large storm near the center of this Voyager image persisted for months, astronomers observing with Hubble know it has since dissipated.

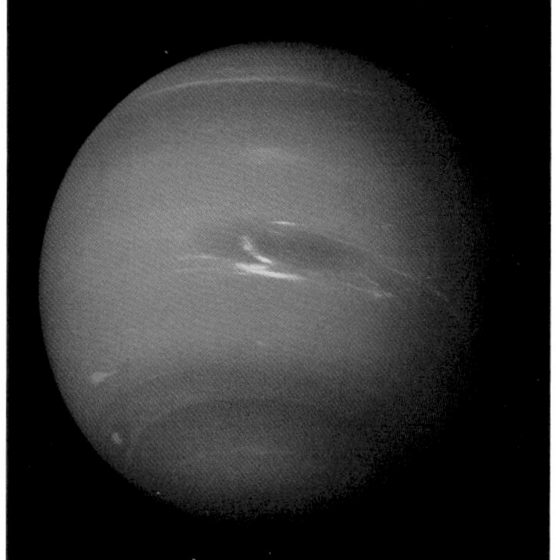

Upcoming Neptune Oppositions		
Date of Opposition	***Peak Magnitude***	***Constellation***
August 17, 2009	7.8	Capricornus
August 20, 2010	7.8	Capricornus
August 22, 2011	7.8	Pisces
August 24, 2012	7.8	Pisces
August 26, 2013	7.8	Pisces
August 29, 2014	7.8	Pisces
August 31, 2015	7.8	Pisces

Asteroids and Dwarf Planets

Once you account for the eight planets, there's not much left in the Sun's family. Although 100,000 or more objects populate the solar system, their masses add up to less than the smallest planet. Astronomers now recognize three dwarf planets: the asteroid Ceres, which orbits the Sun between Mars and Jupiter; Pluto, a major planet before its 2006 demotion; and distant Eris. Of all the small bodies, Ceres and its asteroid siblings are easiest to spot.

A planet until the International Astronomical Union downgraded it in 2006, Pluto now sits uneasily in the category of dwarf planet. Pluto is small and lies far from the Sun. This means it glows feebly at 14th magnitude, and you'll need an 8-inch or larger telescope to spot it. Even then, it looks just like a dim star. Through the eyes of Hubble, however, Pluto (left) and its large moon Charon appear distinct. Pluto also has two other, much fainter moons.

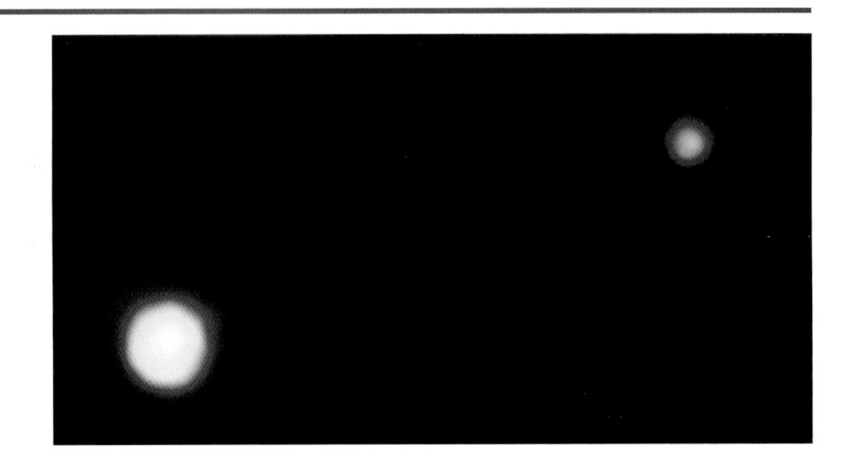

Of all the solar system's asteroids, Vesta glows brightest. At favorable oppositions—the next one comes in August 2011—it reaches naked-eye visibility. Like most asteroids, Vesta resides in the main belt between the orbits of Mars and Jupiter. At 330 miles in diameter, it ranks third in size after Ceres and Pallas. (It appears brighter than those two asteroids because of its more reflective surface.) Scientists used Hubble observations to create this model of Vesta.

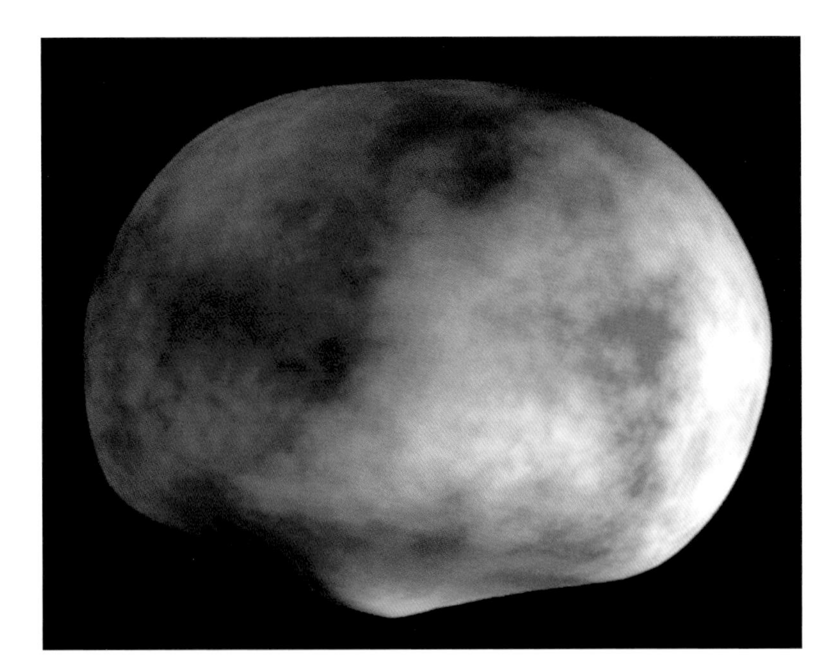

Asteroids have more in common with planets than astronomers once thought. Take this view of the main-belt asteroid Ida, returned by the Galileo spacecraft when it flew past in 1993 on its way to Jupiter. Lurking just to the right of the 35-mile-long asteroid is a tiny moon. The object, named Dactyl, measures barely a mile across and lies only about 60 miles from Ida's center. Since Dactyl's discovery, astronomers have found dozens of other asteroids with moons.

Like most smaller asteroids, Eros looks more like a potato than a sphere. What sets Eros apart is its orbit. Unlike main-belt asteroids, Eros is a "near-Earth object," meaning it occasionally approaches our planet. Objects like Eros have crashed into Earth before, with devastating results, and will again. Astronomers have spent the last decade trying to catalog as many of these near-Earth objects as possible to learn which ones may pose a threat.

FAQ

Is that an asteroid I see?

You've heard it before: Asteroids (and Pluto) look just like faint stars through a telescope. So, how can you tell the difference? Remember that solar system objects move relative to the stars. So, sketch or photograph the star field in which your target lies, and then return to the same field a night or two later. The "star" that has changed position belongs to our solar system.

Comets: Relics of the Solar System's Birth

Vast reservoirs of icy comets exist beyond the planets. Some of them reside in the relatively near Kuiper Belt, while others lie in the distant Oort Cloud. Occasionally, a close encounter with another object will nudge a comet to fall toward the Sun. If the trajectory is right, the Sun's heat will boil off the comet's icy mantle and create a glowing cloud of gas and dust along with a lengthy tail. Such bright comets provide a thrill to all who see them.

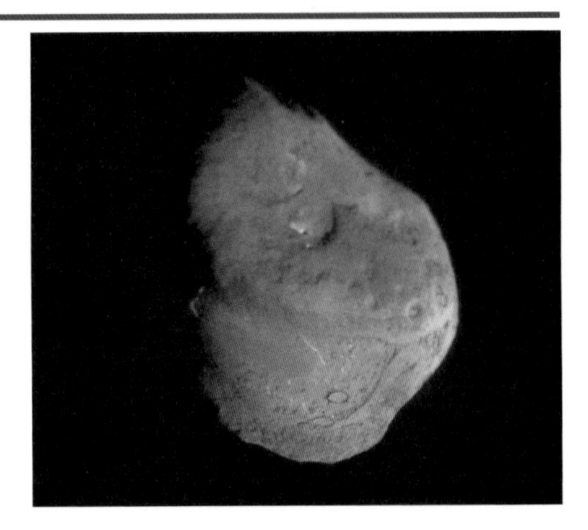

In July 2005, the Deep Impact spacecraft flew past Comet Tempel 1 and returned close-up images of its nucleus. (It also released a heavy projectile that slammed into the nucleus, creating a stadium-sized crater that enabled scientists to learn about the comet's composition and other physical properties.) Astronomers think most comet nuclei look like this: mostly dark with patches of ice that erupt under the Sun's heat. Think of them as "dirty snowballs." This short-exposure composite image didn't capture the erupting jets of gas and dust.

At the same time the Deep Impact probe was visiting Comet Tempel 1 (above), the Hubble Space Telescope also observed the comet. Hubble recorded a vast cloud of gas and dust, called a coma, surrounding the comet's nucleus. This veil always hides the nucleus from earthbound observers. In a particularly active comet, the coma can span a million miles or more. The gases are mostly molecules of water vapor, carbon dioxide, and carbon monoxide.

FAQ

When is the next great comet?

Most of the great comets come with little warning. Often, they're new visitors from the Oort Cloud and have never been seen before. To avoid missing a grand spectacle, it pays to monitor comet-observing Web sites. Two of the best are aerith.net/comet/future-n.html and www.cometchasing.skyhound.com.

A great comet like Hale-Bopp, which graced Earth's sky in 1997, often shows two distinct tails. The bluish tail consists of gases released from the comet's nucleus by the Sun's heat. As the gases rise into the comet's coma, sunlight ionizes the molecules. The solar wind then picks up these charged molecules and blows them away from the Sun.

The yellowish tail consists of dust particles liberated from the comet's nucleus as the gas erupts. These particles simply reflect sunlight, yielding the yellow hue. Solar radiation pushes on the dust, so this tail also points away from the Sun. Strange as it may seem, both comet tails always point away from the Sun. So, when you see a comet heading back toward deep space, the tail lies in front of the comet's path, not behind.

CONTINUED ON NEXT PAGE

The nuclei of most comets aren't held together strongly. A close pass by the Sun or Jupiter can break the comet into pieces. That's what happened to Comet Shoemaker-Levy 9, which split into more than 20 fragments following a close encounter with Jupiter in 1992. Jupiter's gravity not only ripped the comet apart, but also captured the fragments into orbit. In 1994, those fragments smashed into Jupiter's cloud tops.

During a week-long period in July 1994, fragments from Comet Shoemaker-Levy 9 rained down on Jupiter. In this sequence taken by Hubble, the bottom image shows the plume created by the impact of one of the bigger fragments. The next image reveals a large region of dark material dredged up from below by the impact. The next image displays a second dark impact site, while the final image (top) demonstrates the effects on the impact material from Jupiter's strong winds.

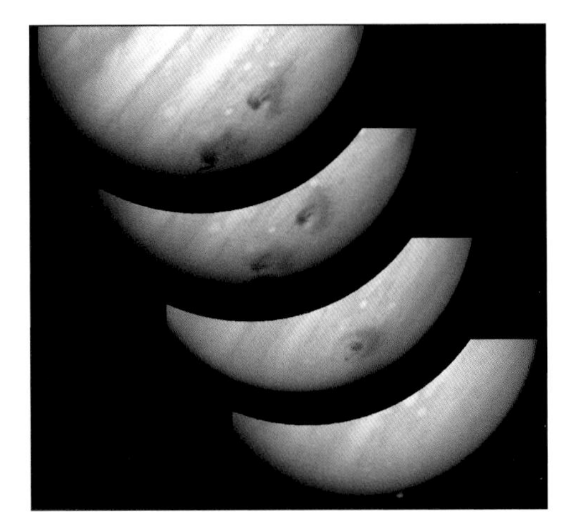

TIP

The Comet-Meteor Connection

All the dust released by a comet during multiple passes by the Sun eventually spreads out along the comet's orbit. If Earth happens to encounter this debris trail, we'll see a meteor shower at the same time each year. We pass through debris from Comet Halley twice each year, creating May's Eta Aquarid shower and October's Orionid shower.

In early 2007, we witnessed the brightest comet in 40 years. Comet C/2006 P1 (McNaught) blazed in Southern Hemisphere skies for several weeks, and briefly grew bright enough to see in daylight. (Unfortunately, it was invisible to northerners at its peak.) Comet McNaught gained fame from its huge dust tail, which spanned 30°.

Like most great comets, both McNaught and Hale-Bopp were making their first trek in from the distant Oort Cloud. So, their nuclei were especially ice-rich, which led to their spectacular displays. Scientists like to study such pristine comets because they haven't changed much since the birth of the solar system 4.65 billion years ago. Examining fresh comets gives researchers an invaluable look back to conditions in the early solar system.

chapter 10

Observe the Deep Sky

When it comes to observing, "deep sky" refers to everything beyond the solar system. From the nearest star beyond the Sun (the triple system known as Alpha Centauri, just 4 light-years distant) to the most distant galaxies billions of light-years away, these deep-sky objects provide endless hours of observing opportunities.

In this chapter, we'll explore the main types of objects visible with the naked eye, binoculars, and small-to-intermediate telescopes, and learn what makes these objects tick. Our cosmic journey will take us from stars and star clusters, through the clouds of gas and dust in our galaxy, and out to the universe of galaxies beyond.

A Universe of Stars

Stars come in all sizes, from dwarfs weighing less than one-tenth what the Sun does to behemoths with 100 times the Sun's mass. Every normal star generates energy in its core, which slowly works its way to the surface. (It can take hundreds of thousands of years for this energy to complete the journey.) Stars are the basic building blocks of galaxies, and the ultimate source for almost all light. They offer a dazzling array of observing opportunities.

Single Stars

In heavyweight stars, like Virgo's Spica (seen here) and most of Orion's luminaries, the weight of overlying matter crushes the gas at its core far more than in Sun-like stars. That pressure raises a star's interior temperature greatly. This temperature rise, in turn, causes fusion reactions to proceed at an accelerated clip. The star ends up with a high luminosity and a surface temperature of several tens of thousands of degrees. Hot stars glow with a blue-white hue, and so, when you see a star this color, you know it's hot.

Stars that appear red have cool surfaces (*cool* meaning a few thousand degrees). Stars get there by one of two paths. Small stars generate little energy in their cores, and so not much reaches their surfaces. High-mass stars, on the other hand, eventually evolve into bloated giant or supergiant stars. Their surfaces, like those of Betelgeuse (near right) and Antares (far right), lie so far from their central heat sources that they, too, glow reddish.

Double and Multiple Stars

You might not guess it by looking at the night sky, but a majority of stars don't live by themselves. Single stars like the Sun are the cosmic exceptions. Double stars typically orbit close enough to each other that they appear as one to the naked eye. But a telescope can often split a double star into its components, frequently revealing contrasting colors. Stars occasionally come together in even greater numbers, as in Lyra's famous "double-double."

The best color contrast for a double star in the northern sky? Most observers would answer Albireo, the star at the base of the Northern Cross in the constellation Cygnus. The brighter star shines with a golden color, while the fainter component appears sapphire blue. You can split Albireo easily with a small telescope at low power.

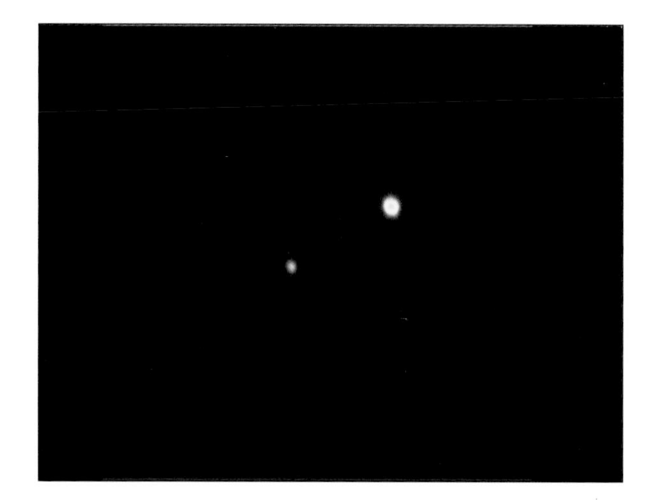

Gaze through a telescope at the double-double—also known as Epsilon (ε) Lyrae—and you won't be amazed with incredible colors. Instead, you'll be enthralled by seeing four related stars in a single field of view. Epsilon is easy to find, located just 1.5° from the brilliant star Vega. With binoculars, you should see two stars. But aim a small telescope at this pair, and each of these stars splits in two. All four stars appear almost equally bright, enhancing the scene.

CONTINUED ON NEXT PAGE

Variable Stars

From here on Earth, stars seem to shine with a constant brightness. Our Sun, after all, puts out nearly the same amount of light that it did 1,000 years ago, and will 1,000 years in the future. At a casual glance, most nighttime stars look steady as well. But a lot of stars vary in brightness. Some vary because they pulsate more-or-less regularly. Others change as orbiting companions block their light. And explosions cause others to vary dramatically.

Delta (δ) Cephei varies by a factor of nearly two—from magnitude 3.6 to 4.3 and back again—every 5.4 days. The change comes about because Delta is nearing the end of its life and is no longer stable. The pulsations that result cause the star to vary in size, temperature, and—most noticeable to us—brightness. You can track these changes by comparing Delta's brightness with those of nearby stars. (The map numbers signify magnitudes with decimal points omitted.)

With a nickname like "The Wonderful," you might think Mira would be conspicuous. More times than not, however, you'll need binoculars to see this autumn star. The red-giant star's brightness varies so much that it often falls below naked-eye visibility. But catch it at the right point in its roughly 11-month period, and it shows up easily. Otherwise, you'll have to break out optical aid. Astronomers expect Mira to peak in November 2009, and every 11 months thereafter.

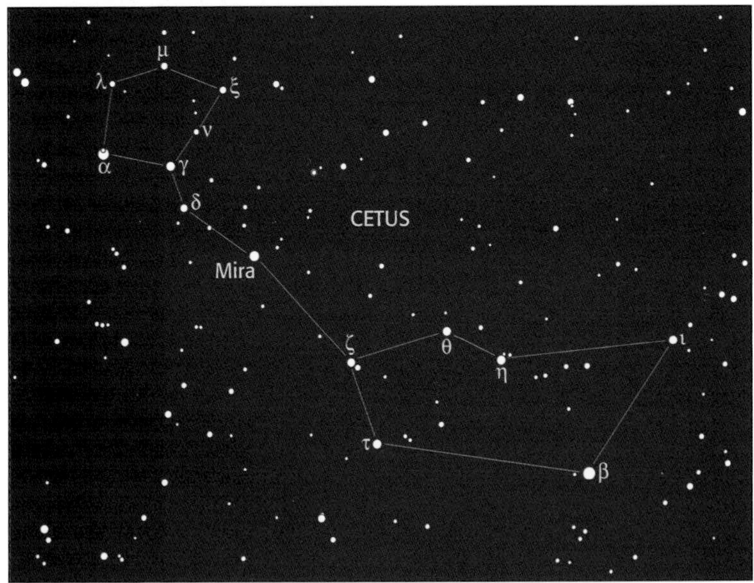

Algol, the "Demon Star," owes its nickname to a rapid and noticeable drop in brightness every 2 days, 20 hours, and 49 minutes. During that period, it goes from magnitude 2.1 to 3.4 and back again. It remains at its faintest for about 10 hours. You can gauge its brightness with this chart, which shows the magnitudes of nearby comparison stars. The decimals have been eliminated to avoid confusion with other stars.

Algol belongs to the class of variable stars known as eclipsing binaries. In these systems, two stars orbit each other along our line of sight from Earth. So, once each orbit, the fainter star passes in front of the brighter one and the system dims noticeably. Half an orbit later, the bright star eclipses the dim one, and the total brightness drops just a bit. At all other times, the light from both stars shines through, and we see the system's typical brightness.

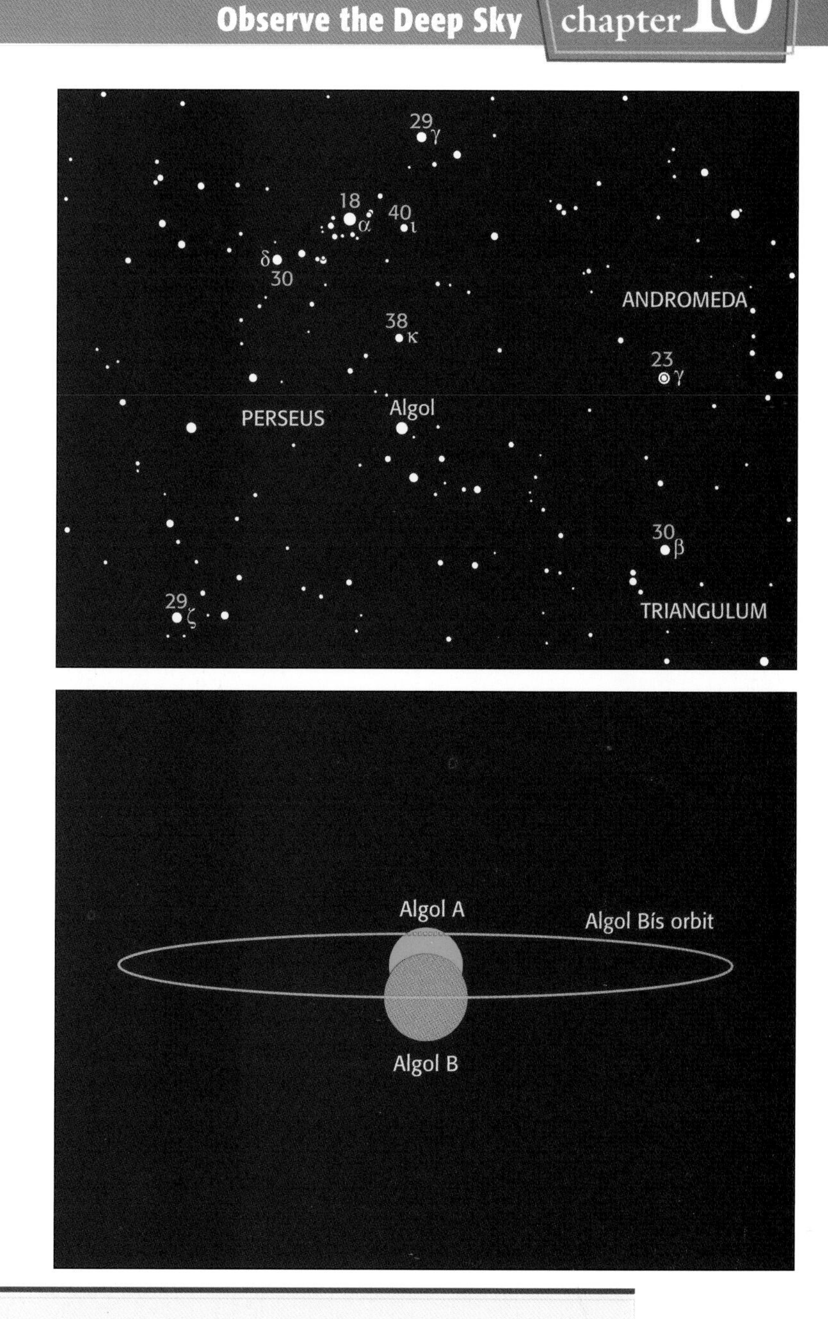

TIP

To Find More Variable Stars

Thousands of variable stars populate the sky. The American Association of Variable Star Observers (AAVSO) is a great organization for people who like to track variables. At its Web site (www.aavso.org), you can learn about variables, get finder charts, and submit observations. Astronomers use these observations to better understand variable stars.

Open Star Clusters

Open star clusters typically contain hundreds to around a thousand stars, and rank among the biggest and brightest deep-sky objects. And for those who observe from the city or suburbs, light pollution doesn't seriously affect the view. These glittering gems usually lie near the band of the Milky Way. You'll see the most in winter and summer, but you can find fine examples any time of year. Because of their size, the best views usually come at low power.

The Beehive star cluster (M44) in Cancer spans three times the Moon's width. At this large size, your best views come through binoculars that deliver 10× or slightly more. The Beehive contains more than 100 stars that combine to glow at 6th magnitude—just bright enough to spot with the naked eye from a dark site. Ancient mariners used this fact to predict the weather. If they couldn't see the cluster on an otherwise clear night, they assumed a storm was coming.

The constellation Scutum lies in the heart of the summer Milky Way and contains one of its star attractions. The pretty open cluster M11 shines bright enough to see without optical aid from a dark site. Small telescopes provide the best views, however, revealing more than 100 stars. Many of those stars are packed tightly at the cluster's center. M11 earned the nickname "Wild Duck Cluster" because its shape reminded one 19th-century observer of a flock of ducks.

Nestled between Cassiopeia and the main body of Perseus are two open clusters perched next to each other. The Double Cluster, cataloged as NGC 869 and NGC 884, looks great through binoculars and ranks among the autumn sky's top deep-sky objects.

Through a telescope at low power, you'll see several blue, yellow, and red gems mixed among the dominant white stars. At closer inspection, you should notice that NGC 869 (the western cluster, which appears at right in the photo) has its stars packed more closely together. Overall, however, NGC 884 contains more bright stars.

Globular Star Clusters

Globular clusters crowd hundreds of thousands of stars into a sphere roughly 100 light-years wide. These giants gather in our galaxy's halo, so you'll find the greatest concentration in Sagittarius, Scorpius, and Ophiuchus. Many observers rate globulars among their favorite objects. Often visible with little or no optical aid, they explode with detail in larger telescopes under dark skies. Most date from the galaxy's infancy more than 10 billion years ago.

Just off the nose of Pegasus the Winged Horse, you'll find the globular cluster M15. Although it's bright enough to see without optical aid under a dark sky, you'll need binoculars to detect the fuzzy shape that defines it as more than a star. A small telescope resolves dozens of cluster members surrounding the bright core. Many of these stars form streamers radiating from M15's center.

Every type of deep-sky object has its preeminent member. For globular clusters, that object is Omega Centauri (NGC 5139). Omega holds a million stars and shines bright enough to see with the naked eye from the suburbs—if you live in the southern United States. Omega's home constellation, Centaurus, lies too far south for northerners to get a good look. If you can see it, both binoculars and telescopes deliver stunning views, revealing up to a thousand stars.

You can find the globular cluster M13 along the western edge of the Keystone asterism in Hercules the Strongman. For most northern observers, this is the finest globular in the sky. (Eight of the ten brightest globulars lie south of the celestial equator.) A 6th-magnitude object, M13 shows up well as a fuzzy "star" to the naked eye under a dark sky. A small telescope turns M13 into a wonder. Through an 8-inch telescope, expect to see 100 or more stars across the globular's face.

Emission Nebulas

The next generation of stars are being born as you read this, in vast clouds of gas and dust known as emission nebulas. These star factories line the spiral arms of our galaxy. They glow from the light of the hot, young stars being forged inside. These stars radiate ultraviolet energy, which excites the surrounding gas and causes it to glow. For backyard observers, emission nebulas are some of the most detailed and beautiful deep-sky objects.

Just north of Sagittarius, in the constellation Serpens, lies one of the best-known nebulas. The Hubble Space Telescope made the Eagle Nebula (M16) famous when it took its portrait in the 1990s. But you need a wider field of view than Hubble's to see the Eagle's entire shape. The Eagle actually combines an emission nebula with an open cluster. As in many star-forming regions, a cluster has already formed in the nebula's interior. A 6-inch telescope shows both nicely.

The only deep-sky object named for one of the United States, the California Nebula (NGC 1499) resides in the constellation Perseus the Hero. The nebula glows bright enough to be glimpsed with the naked eye (through a nebula filter) under excellent conditions. But the light spreads across more than 2° of sky, and so it has a low surface brightness. For those with a telescope, use your lowest magnification and scan across its extent.

Not many constellations look unassuming to the naked eye but explode with deep-sky riches. Monoceros is one of the few. The Rosette Nebula (NGC 2237 through 2239) ranks among the constellation's best. This emission nebula surrounds the naked-eye star cluster NGC 2244. The nebula spreads out enough that you'll want to use a fairly large aperture and low magnification to see it well. A nebula filter will dim the dozens of cluster stars and make the nebula stand out.

TIP

Try a Filter for Greater Clarity

A nebula filter will give you better views of most emission nebulas. "Deep-sky," "LPR" (light pollution reduction), and "UHC" (ultra-high contrast) filters work by blocking unwanted light and letting through a nebula's light. The increased contrast helps you to see nebular details. You'll notice a bigger benefit with larger scopes, which deliver more light to begin with.

Dark and Reflection Nebulas

While hot gas causes emission nebulas to glow, cold gas and dust reflect or block light. Sometimes, thick dust bands create fanciful silhouettes by blotting out more distant starlight. At other times, the gas and dust take on an eerie bluish glow as they reflect light from stars located nearby. In nearly every case, however, they offer a subtle reminder of the universe's capacity to make visual splendors from the ingredients at hand.

When energetic starlight hits gas atoms, it causes the gas to glow with a characteristic reddish color. But if a star lies too far from a cloud of cold gas and dust, or just isn't blazing hot, its light reflects or scatters off the material. The tiny particles in the cloud give it a blue color, for the same reason Earth's daytime sky appears blue. An example of this effect is the Iris Nebula (NGC 7023) in Cepheus, which appears as a bright, circular haze in modest-sized telescopes.

Few people could look at this dark mass and not see a horse's head. A dark nebula, the Horsehead Nebula lies in Orion, just south of the bright star at the eastern end of Orion's belt. It stands in front of and blocks the light from the emission nebula IC 434. The Horsehead can be glimpsed in a 5-inch telescope, but it's a challenge to most observers even with much larger scopes. Use the largest scope available, get totally dark-adapted, and give it a shot.

Appropriately enough, the Snake Nebula belongs to the constellation Ophiuchus the Serpent-bearer. Here, a dusty cloud blocks the more distant stars. The cloud meanders against the rich Milky Way background. The best views of the Snake come through binoculars or a telescope that provides a wide field. Several other dark nebulas lie in this neighborhood, and so it's a good region to scan with binoculars.

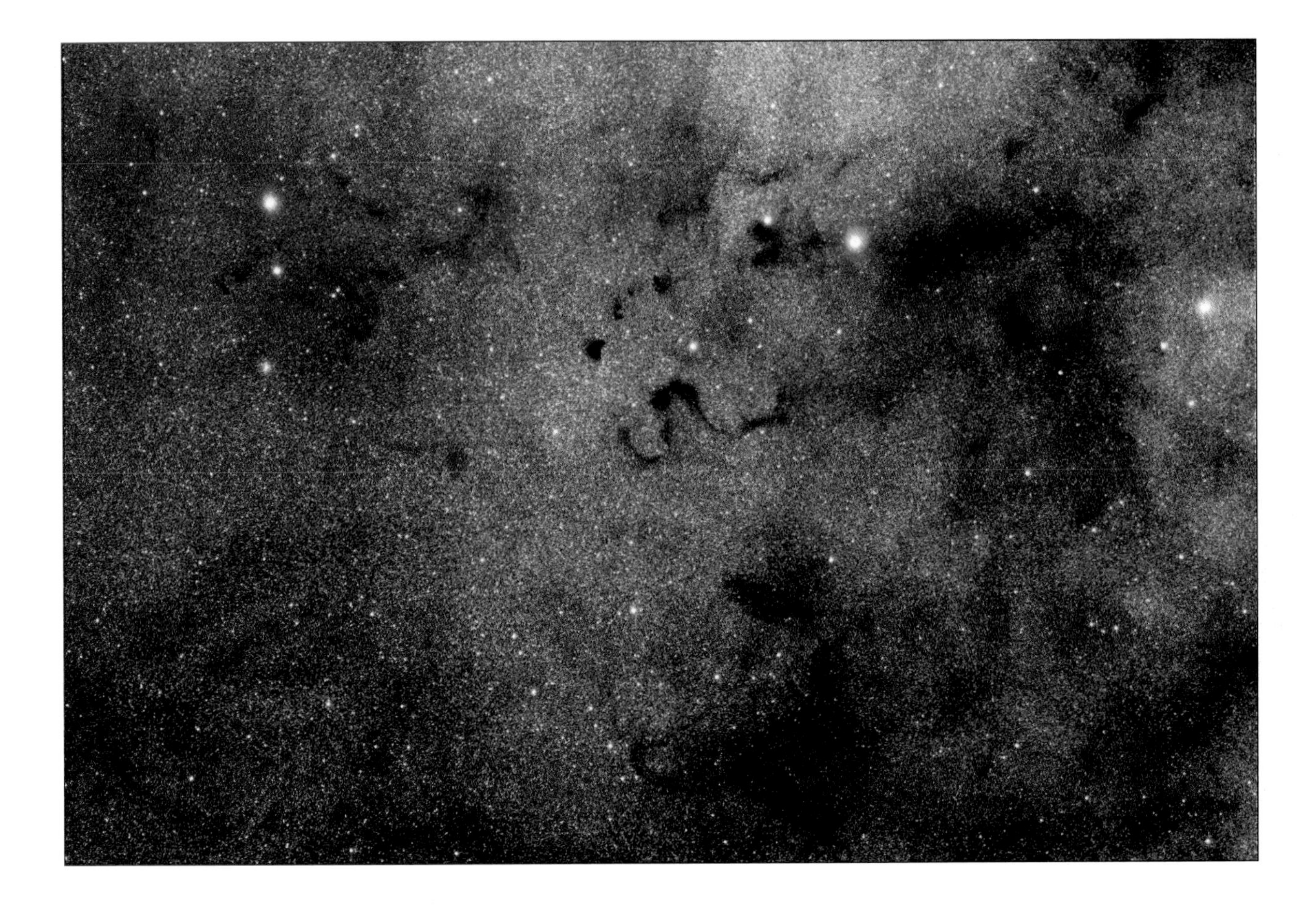

TIP

Leave Your Nebula Filter at Home for These Objects

Despite the name, a nebula filter usually won't help you see a dark or reflection nebula. Because these objects usually reflect starlight or block more distant starlight, a filter that prevents most starlight from passing through will reduce contrast and make the nebulas harder to see. The one exception: the Horsehead Nebula, which appears more distinct through a filter because it enhances the emission nebula behind it.

Planetary Nebulas

When a Sun-like star nears the end of its life, it becomes unstable and "puffs off" its outer layers. Often, this process happens several times. After the star evolves into a white dwarf, its high-energy radiation causes these gas layers to glow. Planetary nebulas—so named because early observers thought they resembled Uranus in the eyepiece—have some of the most complicated and beautiful shapes in the cosmos.

The Dumbbell Nebula (M27) in Vulpecula takes its name from two symmetric lobes that define its shape. The Dumbbell ranks among the biggest and brightest planetaries because it lies fairly close to Earth (some 800 light-years away). Binoculars easily pick up M27's bright glow. A small telescope shows the two lobes and several foreground stars. Because of the Dumbbell's high surface brightness, you can increase the magnification as much as conditions permit.

Some planetary nebulas look pretty simple, while others show surprising complexity. The Cat's Eye Nebula (NGC 6543) in Draco certainly falls into the latter category. The Cat's Eye shows at least 11 concentric shells surrounding the central white-dwarf star, each apparently expelled from the dying star during a separate event. An 8-inch telescope clearly reveals the nebula's blue-green color, and a hint of structure around the white dwarf.

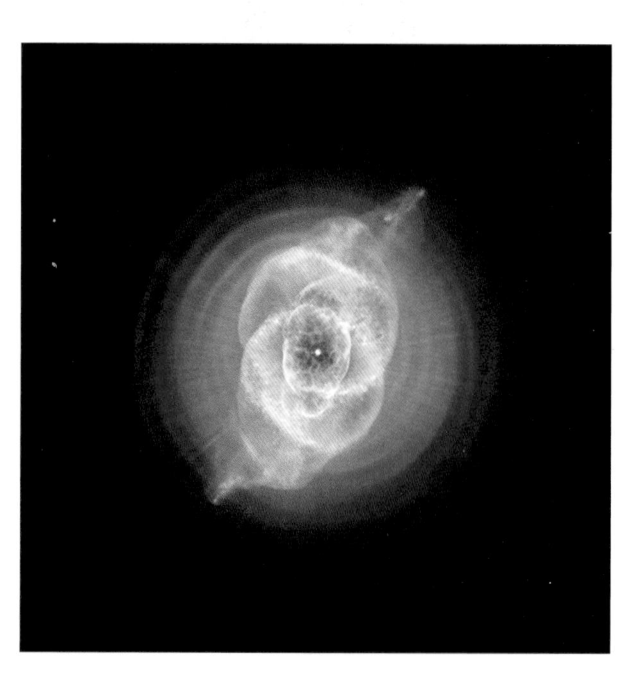

Few deep-sky objects are as easy to find as the Ring Nebula (M57). It lies midway between the 3rd-magnitude stars Beta (β) and Gamma (γ) Lyrae, the two stars at the bottom of the parallelogram southeast of Vega. The Ring looks just like its name, and can be viewed through a small telescope. You'll see a small, bright oval that's remarkably clear inside. As with most high-surface-brightness objects, you can crank up the power to see detail on M57.

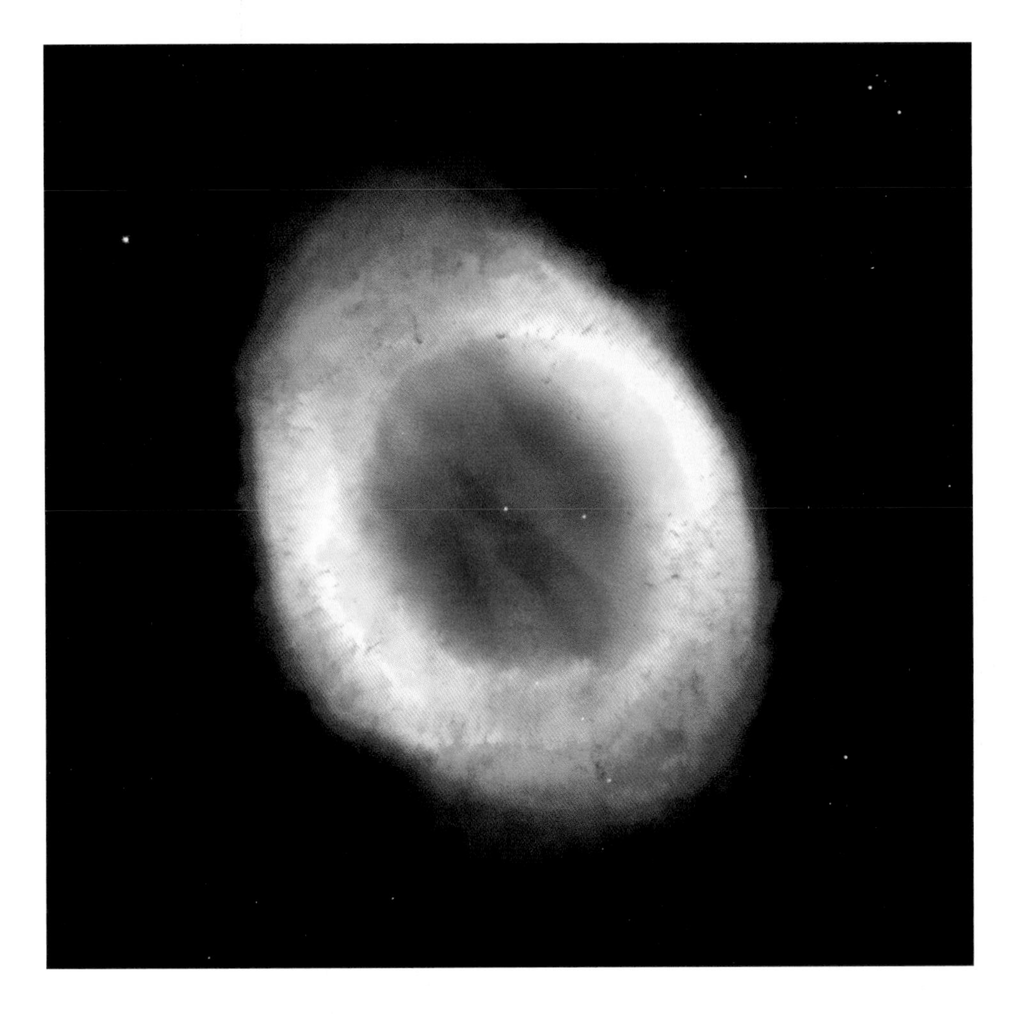

TIP

Use an Oxygen III Filter on Planetary Nebulas

Most planetary nebulas benefit from using a filter that passes only light at the wavelength of doubly ionized oxygen (OIII). Planetaries emit much of their light at this wavelength, and so OIII filters will dim almost everything else in the field of view and let the nebula stand out.

Supernovas and Their Remnants

Most stars, including the Sun, will die peacefully as planetary nebulas. The most massive stars aren't as lucky. Stars born weighing more than 12 times what the Sun does can't settle down in old age. When they run out of fuel, they self-destruct and create titanic explosions. A supernova can shine with the light of billions of normal stars and be viewed across the cosmos. And it leaves behind a chaotic mess worth observing from Earth.

In 2004, a massive star in the galaxy NGC 2403 reached the end of its life. The star exploded, becoming as bright as 200 million Suns. In this Hubble Space Telescope image, the supernova (arrow) rivals some stars in our own galaxy, despite lying 11 million light-years farther away. Such explosions happen roughly once every 100 years in a typical spiral galaxy. No one knows when the next will occur, and so it pays to keep an eye on the sky.

Humans haven't seen a supernova in our galaxy in more than 400 years. We appear long overdue, although it's possible one occurred behind thick clouds of dust and went unnoticed. Among nearby stars, 6th-magnitude Eta Carinae seems a prime supernova candidate. This star, immersed in a dusty nebula, weighs at least 100 times what the Sun does and almost certainly will become a supernova someday. The day of reckoning could be thousands of years in the future.

The Pencil Nebula (NGC 2736) represents a tiny part of the vast Vela supernova remnant. The entire remnant spans more than 100 light-years, while this part measures less than a light-year across. Although some supernova remnants, like the Crab Nebula in Taurus and the Veil Nebula in Cygnus, shine quite brightly, the Vela remnant is harder to see. The best views come in 12-inch or larger telescopes at low power while using a nebula filter.

Spiral Galaxies

If stars are the building blocks of galaxies, then galaxies are the building blocks of the universe. More than 100 billion galaxies populate the cosmos, and they come in all shapes and sizes. None offers more visual appeal than spirals. These pinwheel-shaped objects can appear edge-on, face-on, or at any angle in between. Some even have a bar of gas, dust, and stars running through their centers.

The finest edge-on spiral galaxy in the sky, NGC 4565, lies in Coma Berenices. This spring object has the shape of a typical edge-on spiral, with a long, thin dust lane running lengthwise. In NGC 4565's case, the lane shows up best in front of the galaxy's small central bulge. An 8-inch telescope will show the disk stretching five to six times longer than the bulge's breadth. Bigger telescopes extend the galaxy's apparent length.

In a majority of spiral galaxies, the arms spiral all the way into the center. But a significant fraction of spirals show the arms winding out from the end of a long bar of stars, gas, and dust. Astronomers now know the Milky Way is a barred spiral. The classic barred spiral NGC 1300, seen here, glows at 10th magnitude in the constellation Eridanus, and appears best on winter evenings. Through a telescope, the arms glow brightest where they connect to the bar.

Although the Whirlpool Galaxy (M51) lies in the spring sky, it's not just another spring galaxy. The Whirlpool stands apart as a spectacular face-on spiral. You can find it in Canes Venatici, just south of its border with Ursa Major. An 8-inch telescope shows the spiral arms nicely, and larger scopes show fine detail. Interestingly, M51 is interacting with the small galaxy NGC 5195 (which appears as a blob at the tip of one spiral arm). The collision hasn't had much noticeable impact yet.

Elliptical Galaxies

While spiral galaxies offer variety, elliptical galaxies look pretty much alike. The main difference is in how elliptical they look. Ellipticals show less variety in appearance and more in size. The universe's biggest galaxies, containing a trillion or more stars, are ellipticals. So are the smallest galaxies, with star counts only in the millions. Yet even among the largest galaxies, you can find intriguing ellipticals if you know where to look.

The massive elliptical galaxy NGC 4881 appears to be close to a sphere. As with many large ellipticals, NGC 4881 controls a large family of globular star clusters. These appear as points of light surrounding the galaxy. NGC 4881 lies hundreds of millions of light-years from Earth, in the vast Coma Cluster of galaxies. You'll need an 8-inch telescope to make out the galaxy's 13th-magnitude glow.

The elongated elliptical galaxy NGC 1132 looks more like a cigar than a sphere. This winter object lies in Eridanus at a distance of roughly 300 million light-years. Astronomers suspect that many giant ellipticals, including NGC 1132, grew by swallowing a lot of smaller galaxies. (Such galactic cannibalism goes on in many large galaxies, including the Milky Way.) You'll need a 10-inch or larger telescope to spy the light from 14th-magnitude NGC 1132.

Not all elliptical galaxies are boring. NGC 5128, also known as Centaurus A because it's a bright radio source, lies just 10 million light-years from Earth and shows a lot of detail. Most interesting is the dust band. Almost all ellipticals stopped making stars long ago and no longer possess the gas and dust essential for new stars. In NGC 5128's case, the dust comes from a small spiral galaxy that recently merged with the main object. The merger triggered star birth in the companion spiral. Although Centaurus A lies low in the sky for northern observers, a small telescope shows it as round with a wide dust lane dividing it in two.

Irregular and Peculiar Galaxies

The strangest-looking galaxies—and among the most visually arresting—are those lacking symmetry. Some of these galaxies likely were born strange, while others faced massive changes triggered by collisions. Whatever the cause, most of these oddballs are worth observing through a telescope.

Deep in Earth's southern sky lies a galaxy first seen by Europeans during Magellan's circuit of the globe. The Small Magellanic Cloud (SMC) lies just 210,000 light-years from Earth and shows incredible detail through binoculars or any size telescope. If you ever visit the Southern Hemisphere, be sure to observe this irregular galaxy. The bright object to the upper right of the SMC is the globular cluster 47 Tucanae (NGC 104), the second-brightest globular in the sky.

The Large Magellanic Cloud (LMC) ranks as the biggest galaxy visible from Earth. Unfortunately, like the SMC, the LMC lies deep in the southern sky. This irregular galaxy shows great detail with just the naked eye, with binoculars, or through any telescope. You could spend a year observing all the wonders it offers. But the most exceptional object in the LMC is the Tarantula Nebula (NGC 2070), the largest known star-forming region in the universe.

The Antennae (NGC 4038 and NGC 4039) in Corvus appear peculiar because they are colliding and thus bear little resemblance to a normal galaxy. The two spirals started to interact a few hundred million years ago. Now, brilliant, blue star clusters and reddish emission nebulas dot the object as the merger compresses material and forms new stars. You can see this pair through a 6-inch telescope, although the two objects appear indistinct. Double the aperture and you'll see both of the still-intact galactic cores.

Galaxy Groups and Clusters

Galaxies apparently don't like to be alone. Most galaxies belong to small groups, like our own Local Group of a few dozen galaxies, or to clusters of up to thousands of galaxies. The closest large cluster is the Virgo Cluster, discussed in Chapter 5. But plenty of others exist across the sky for your viewing pleasure. The most appealing clusters tend to be those that show variety among their galaxies or some interactions.

Viewing Seyfert's Sextet, a compact galaxy group, is a challenge. You'll need a 12-inch telescope just to spy a faint glow. But this photo shows what a compact galaxy group looks like. First, the sextet should be called a quartet. The "galaxy" at far right is really a tidal tail pulled from another galaxy, and the small face-on spiral lies much farther away. The other four span 100,000 light-years total, about the same as the Milky Way by itself. One day, these four may merge into one galaxy.

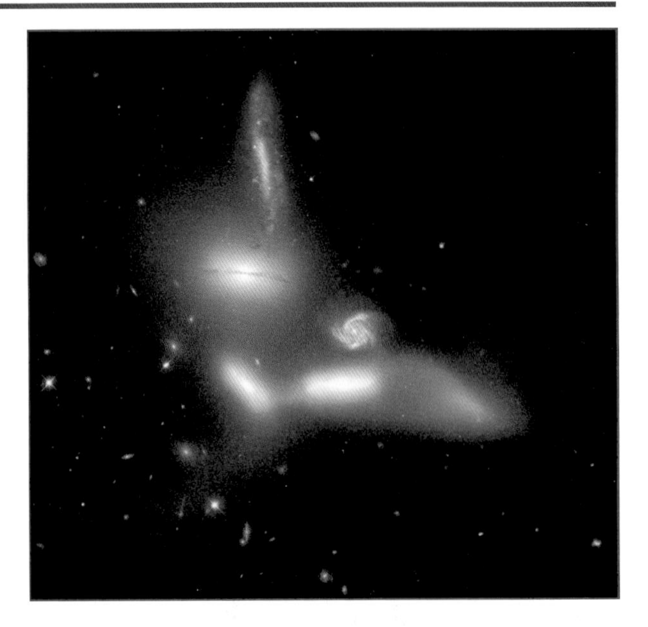

Stephan's Quintet in Pegasus, like Seyfert's Sextet above, shows that collisions and interactions will take place whenever you pack enough galaxies into a limited space. The long tidal tails illustrate what happens when gravity acts on a galaxy's stars. This compact group shines bright enough to view through a 6-inch telescope. Expect to see a small, clumpy glow. A 12-inch telescope resolves the individual galaxies, although you'll be hard-pressed to see details.

Perhaps nowhere else will you find the variety of spiral galaxies seen in the Hercules galaxy cluster. In most rich clusters, ellipticals dominate because collisions tend to destroy the more-delicate spirals. But in the Hercules cluster, plenty of spirals remain with their stores of gas and dust. The cluster, which lies 500 million light-years from Earth, holds 20 or so galaxies brighter than 15th magnitude. So, you should be able to make out the brighter members with a 12-inch telescope. Good luck—this is probably the toughest observing challenge in this book.

chapter 11

Record the Sky

Who among us hasn't gazed on a beautiful twilit sky and wanted to capture the memory? Or viewed a rising Full Moon and thought, "That would make a great picture?" Yet many people think the demands of photographing the night sky lie beyond their equipment or abilities. For many of the scenes you might want to capture, this just isn't true.

Photography isn't the only way to capture what you see. You can create a lasting record of the image produced by your telescope with a sketch. This technique has an added bonus: By studying what you see closely enough to produce an accurate sketch, you're also training your eye to see detail. You'll be amazed at how much more you'll see once you've sketched a few objects.

Sketch the Sky

You don't have to be an artist to sketch what you see through a telescope. All you need is an eye for detail and the patience to practice your technique. Sketches can form a valuable log of your observations. They also help train your eye to see critical detail. A word of advice: Don't expect a deep-sky sketch to match a photograph. Time-exposure photos reveal color and detail in faint objects that the human eye simply can't register.

The Sun has to be one of the easiest solar system subjects to sketch. Keep in mind these two keys for drawing our star. First, save your eyesight by following safe observing practices (see Chapter 8). A solar projection setup like the one shown here lets you view safely and see solar detail. Second, wait until the Sun enters an active phase with at least some sunspots. A sketch without spots will look boring regardless of your skills.

On August 13, 2004, the active region 0656 featured a prominent sunspot group. This group darkened a big enough area of the Sun's disk that observers on Earth looking through solar filters could see it with their naked eyes. This sketch, made by projecting the Sun's image onto white paper, reveals the group's structure. Note the dark umbra of each spot and the surrounding penumbra. Many of these penumbras overlap. Several smaller spots also dot the Sun's surface.

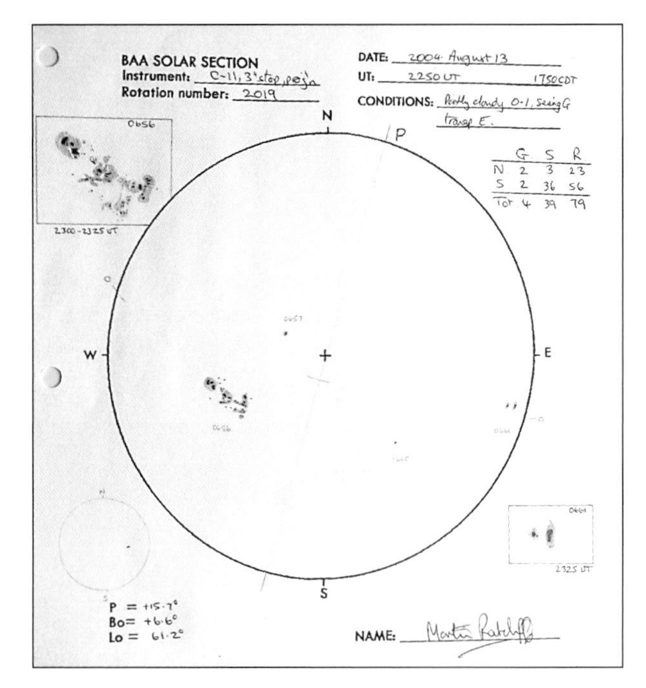

To sketch a deep-sky object, get a sheet of white paper and, with a pencil, draw a circle to represent the telescope's field of view. Make the circle at least three inches in diameter. At the telescope, use a dim red light to illuminate the paper without ruining your dark adaptation. Start the sketch by drawing stars. Begin with the brightest ones, and then fill in fainter ones by triangulating their positions. Use bigger dots to represent brighter stars.

If you're sketching a nebula or galaxy, you next need to lay down a layer of graphite. (Almost all deep-sky sketches are done as negatives, with bright areas represented by dark pencil markings.) Do this by gently rubbing the pencil's edge on the paper. (Eberhard Faber Ebony pencils, available at art-and-craft stores, are great for this.) Then, using a dry or slightly wet finger, lightly smear the graphite so it resembles the shape of the object's diffuse light.

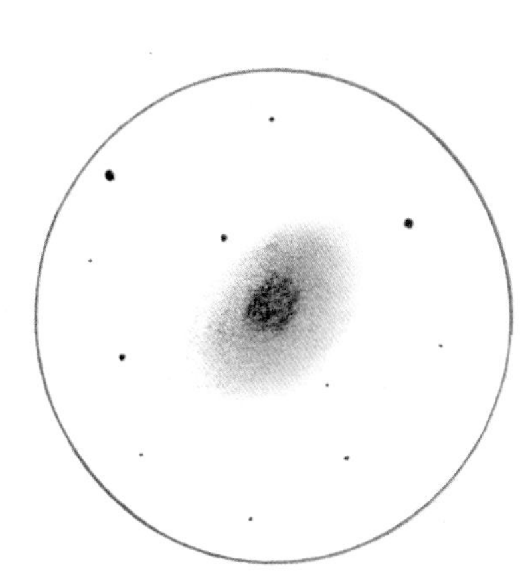

Finish your sketch by smearing, erasing, or adding to the graphite you've already laid down. The goal is to manipulate the sketch so it comes as close as possible to matching what you see in the eyepiece. A good sketch can take anywhere from 5 to 30 minutes to complete. Double stars and open clusters are the easiest targets. Globular clusters tax your abilities a little more, while galaxies and nebulas prove hardest to capture.

CONTINUED ON NEXT PAGE

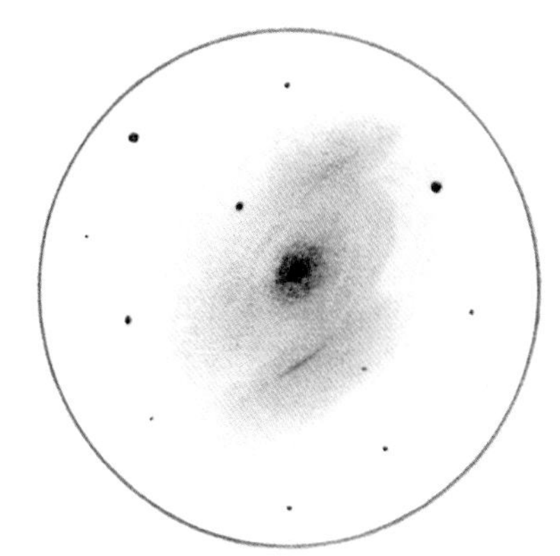

Sketch the Sky *(continued)*

The open star cluster M37 in Auriga makes a good target for beginning sketch artists. Recording the cluster's several dozen stars will exercise your talents at proper star positioning.

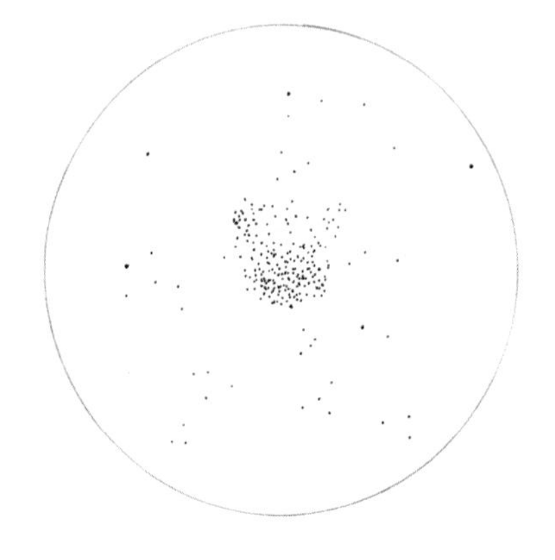

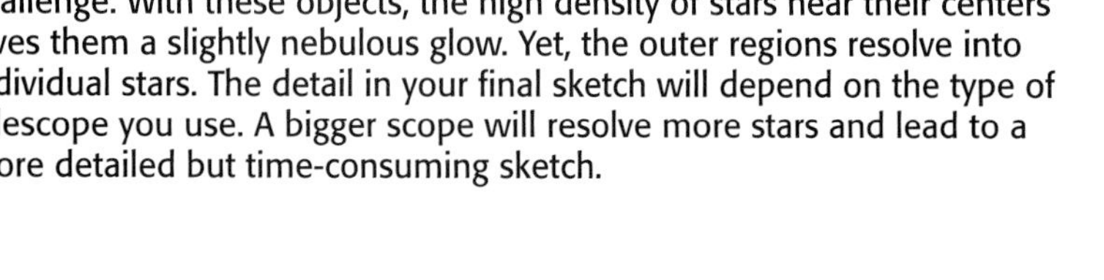

Globular star clusters like the Hercules Cluster (M13) provide more of a challenge. With these objects, the high density of stars near their centers gives them a slightly nebulous glow. Yet, the outer regions resolve into individual stars. The detail in your final sketch will depend on the type of telescope you use. A bigger scope will resolve more stars and lead to a more detailed but time-consuming sketch.

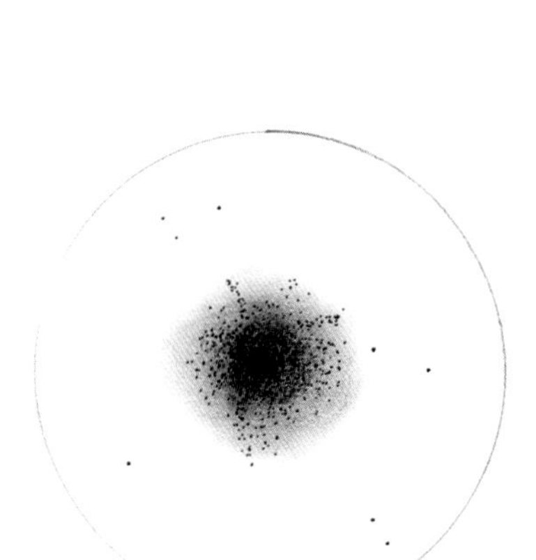

Few emission nebulas possess the complexity of the Orion Nebula (M42). Note in this sketch how the artist had to erase some graphite to represent the dark dust intruding toward the nebula's center. The four stars recorded at the heart of M42 are known as the Trapezium. One of these hot, young stars provides nearly all the radiation that keeps the nebula glowing.

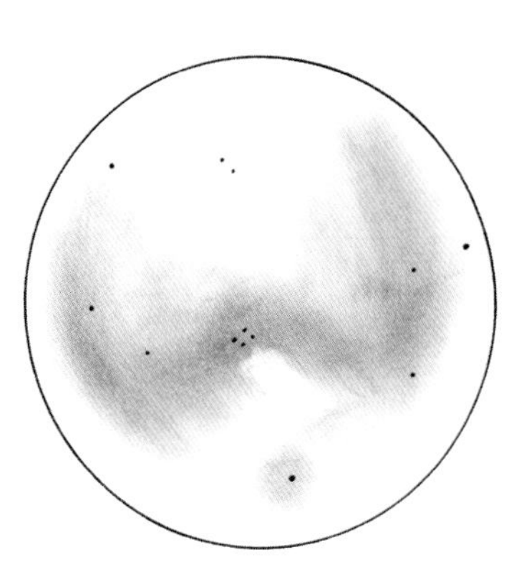

The nearly edge-on spiral galaxy NGC 253 lies in the constellation Sculptor. Also known as the Silver Coin Galaxy, this object ranks among the brightest galaxies in the sky. NGC 253 belongs to the Sculptor Group, one of the closest collections of galaxies beyond our Local Group.

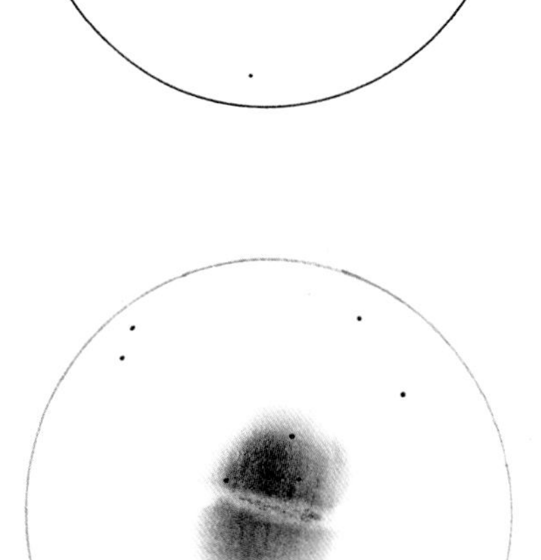

Centaurus A, also known as NGC 5128, is a nearby giant elliptical galaxy in the process of swallowing a smaller spiral galaxy. The dust from the spiral forms a band across Centaurus A's central regions that blocks the light from beyond. In this negative sketch, the dark band appears lighter than the rest of the galaxy.

Ursa Major holds two galaxies that appear in the same low-power telescopic field. The spiral galaxy M81, the bigger and brighter of the pair, lies at the bottom of this sketch. The irregular galaxy M82 appears as a needle at the sketch's top.

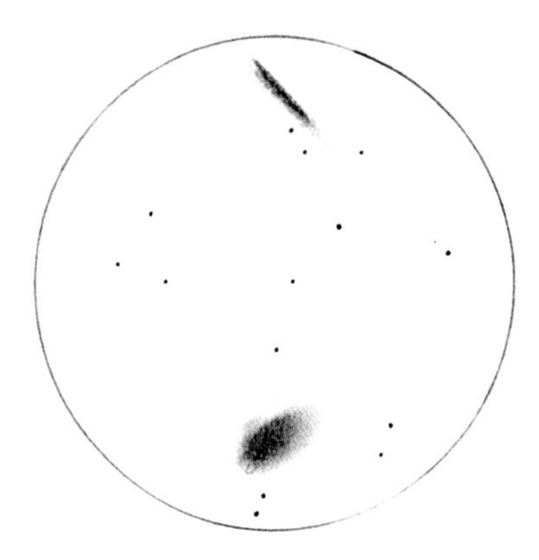

Photograph the Sky

Do you think you need fancy equipment to capture the sky? Think again. All the pictures you see in this chapter's remaining pages were shot with simple camera-on-tripod setups. The key is having a camera that can take time-exposure photos, which will capture the fainter light most astrophotographers seek. Preparation and artistic sensibility count more in creating exceptional portraits than equipment does.

Flip through the pages of this book, and you'll see scores of beautiful photographs. Some were taken through large telescopes, either on the ground or in space. Others were captured through fairly typical backyard telescopes, but using sophisticated techniques and image-processing abilities that can take years to acquire. (Many of the deep-sky objects and planet close-ups fall into this category.)

Yet a large number of these stunning photographs required nothing more than a camera and a tripod. This technique excels at capturing wide-angle views of the heavens that closely match what the human eye sees. Shots of a colorful twilight sky, the Moon hanging above a pretty landscape, a conjunction between two planets, or a vivid auroral display all look best when caught with a fair amount of surrounding sky.

Here's the short list of what you need to shoot the sky: a camera—either digital or film will work—with the ability to take time exposures, a sturdy tripod, and a cable release or electronic remote control to keep the camera from shaking when you press the shutter. If you're photographing under a dark sky, you'll also want a red flashlight to see what you're doing without destroying your night vision.

To capture a twilight scene—and this includes most wide-angle Moon photos and images of pretty conjunctions—set the focus to infinity and let your camera's light meter do its job. If you're using film, bracket the exposures to make sure at least one gets the lighting just right. With a digital camera, you can get immediate feedback and adjust on the fly.

Remember that the sky event is often just one aspect of what you're recording. Plan your shots with an interesting foreground, and the landscape will add to the scene. Think the way you would if you were taking a daylight picture—you don't want telephone wires or something equally ugly to mar the scene. It's much better to have a photogenic building, tree, or landscape in the final shot.

Shooting an aurora, or a lunar or solar eclipse, isn't much different. Again, focus on infinity and let your camera get the exposure in the right ballpark. With auroras, pay attention to how fast the display changes. A long-exposure photo of a rapidly moving formation will show little detail. Opt for short exposures on bright, active displays. During the partial phases of a solar eclipse, don't look through the viewfinder unless you have a proper solar filter in place. During totality, remove the filter and shoot away.

Capturing constellations provides a good introduction to deep-sky astrophotography. For the best results, set the f-stop to f-2 or f-2.8. If you want stars to appear as pinpoints, keep the exposure reasonably short. With a 50mm lens, don't shoot longer than about 15 seconds, or you'll start to see trailing. With a 28mm lens, you can double the exposure before seeing trails.

Not everyone wants to avoid trailing stars. Some of the most dramatic astrophotos record stars arcing across the sky, forming concentric circles around the north or south celestial pole. Film still rules this form of astrophotography. Just be sure your camera can keep its shutter open for a long time without draining the batteries—many star-trail photos last an hour or more.

And, finally, experiment with your technique. None of the images on these pages were taken by photographers on their first attempt. In many cases, careful planning went into the shot. Scout out possible photo locations when you read about an upcoming sky event. Keep abreast of the latest celestial happenings. And the next time you witness a great event, capture the memory.

A great way to capture the sky is to put your camera on a tripod, attach a cable release, and shoot away. It helps when you have something dramatic to photograph, like this green and purple aurora.

CONTINUED ON NEXT PAGE

There's an old saying that goes: "Red sky at night, sailor's delight." It's also not a bad adage for photographers wanting to capture a pretty twilight scene. Watch for nature to supply the scenery, then grab your camera and tripod and get to work.

When the sky's clear, twilight photos often need a focus to look special. In this instance, a slender crescent Moon provides the main focal point, while Saturn appears as a fainter point of light just above the image's center.

CONTINUED ON NEXT PAGE

Your choice of a foreground subject makes a big difference in how the final photograph will look. Here we see Venus and Jupiter (Venus shines brighter and lies to Jupiter's lower right) above an Iranian mosque.

Venus and Jupiter also appear next to each other in this scene (again, Venus is brighter, but this time it lies to Jupiter's upper left). Here, the photographer chose a calm lake and stately tree as a foreground. This photo was taken in deeper twilight than the one opposite, so the sky appears darker and more stars show up.

CONTINUED ON NEXT PAGE

Photograph the Sky *(continued)*

Often in astrophotography, preparation is more important than equipment. It took photographer Anthony Ayiomamitis more than a year to capture the Full Moon rising next to the Parthenon in central Athens. He had to calculate both the Moon's precise position relative to the building and the Sun's distance below the horizon to make sure twilight set the building aglow.

A Full Moon rises above this cityscape during the partial lunar eclipse of May 24, 1994. Photographer Alister Ling created this composite by taking multiple exposures of the rising Moon.

CONTINUED ON NEXT PAGE

The Alaska pipeline gleams from the light of a passing car, but it only serves to add contrast to the greenish aurora glowing above. Painting nearby objects with light (a flashlight will do) is a handy way to make the foreground stand out on an otherwise dark night.

This 5-hour time exposure reveals star trails around the north celestial pole. The bright star with the shortest arc near the center is Polaris. Star-trail photos like this graphically show circumpolar stars, those that never rise or set but instead endlessly circle the pole. From this location in Greece, any star within 37° of the pole is circumpolar.

Appendix

Can't remember the difference between apparent and absolute magnitude? Or how about the locations of the photosphere, chromosphere, and corona in the Sun's atmosphere? A quick glance in the glossary will jog your memory. After the glossary, you will find a list of all 88 of the sky's constellations, from prominent groups such as Orion the Hunter to obscure collections such as Camelopardalis the Giraffe. Another table records the 25 brightest stars visible from Earth. If you need to know whether winter's Aldebaran is brighter than summer's Antares, this is the place to look. If you're in the market for a new telescope, the manufacturers we've compiled in "Telescope Resources" are a great place to start. Finally, if you want to know who took the great photos seen throughout this book, we've cataloged the contributors in "Art Credits."

Glossary

aberration Any of various defects in an optical system that adversely affect an image's quality.

absolute magnitude The magnitude of a celestial object at a standard distance of 32.6 light-years.

achromatic lens A two-element lens that brings light of two colors to a common focus.

active galaxy A galaxy whose central region appears abnormally bright, usually the result of matter feeding a black hole.

albedo The fraction of incoming light (from the Sun, for example) that an object reflects.

altazimuth mounting A telescope mounting where one axis swings horizontally around the horizon and the other moves vertically.

altitude The angular distance of an object above the horizon.

angular diameter The apparent size of an object, measured as an angle. The Moon's angular diameter averages 0.5°.

angular distance The apparent separation between two celestial objects.

annular solar eclipse An eclipse of the Sun during which the Moon passes across the center of the Sun's disk, but appears too small to block the whole Sun. A ring of sunlight remains visible around the Moon.

aperture The diameter of a telescope's main lens or mirror.

aphelion The point where an object orbiting the Sun lies farthest from our star.

apochromatic lens A lens with three or more elements that focuses nearly all colors to the same point.

apogee The point where an object orbiting Earth (like the Moon) lies farthest from our planet.

apparent magnitude A measure of the brightness of an object as seen from Earth.

apparition The length of time, usually measured in weeks or months, that a celestial object remains visible.

arcminute An angle equal to 1/60 of a degree.

arcsecond An angle equal to 1/60 of an arcminute, or 1/3,600 of a degree.

association A large, loose group of young stars.

asterism A recognizable pattern of stars.

asteroid A rocky object, smaller than a planet, that orbits the Sun.

asteroid belt The region between the orbits of Mars and Jupiter where most asteroids reside.

astronomical twilight The level of illumination experienced when the Sun lies between 12° and 18° below the horizon.

astronomical unit (AU) The average distance between the Sun and Earth (approximately 92.9 million miles); typically used to measure distances within the solar system.

aurora A glow high in the atmosphere created when charged solar particles collide with atoms and molecules of air.

azimuth The angular distance, measured eastward along the horizon from due north, to a point directly below an object.

Baily's beads During a total solar eclipse, the bright points of sunlight seen along the Moon's limb just before and after totality.

Barlow lens A lens that increases the effective focal length of a telescope, and thus the magnification of any eyepiece, usually between 2 and 5 times.

barred spiral galaxy A spiral galaxy in which the arms start at the end of a bar composed of stars, gas, and dust.

Big Bang The moment when the universe came into existence, approximately 13.7 billion years ago.

binary star Two stars that are in orbit around each other.

binoculars An optical device consisting of two small, side-by-side telescopes joined together into a single unit.

black hole An object whose gravity is so strong, not even light can escape from it. They range from a few times the Sun's mass to billions of solar masses.

Cassegrain telescope A telescope design in which light collected by the primary mirror bounces to a secondary mirror, which redirects the light to a focus through a hole in the primary.

Cassini Division A large gap in Saturn's rings, usually visible through a small telescope.

catadioptric telescope A compound telescope design that uses elements of both refracting and reflecting telescopes.

celestial equator The projection of Earth's equator onto the celestial sphere.

celestial pole The projection of Earth's axis onto the celestial sphere.

celestial sphere An infinitely large, imaginary sphere centered on Earth. All celestial objects appear to lie on the celestial sphere.

Cepheid variable A type of pulsating variable star whose period of variation reveals its intrinsic brightness.

chromatic aberration An optical aberration in which light of different colors focuses at different distances from the primary lens.

chromosphere A thin layer of the Sun's atmosphere just above the photosphere.

circumpolar star A star that lies close enough to a celestial pole that it never rises or sets. The higher your latitude, the more circumpolar stars you'll have.

civil twilight The level of illumination experienced when the Sun lies less than 6° below the horizon.

coma An expansive cloud of gas and dust that surrounds a comet's nucleus after the Sun's heat starts to vaporize the comet's ices.

comet A small object of ice and rock orbiting the Sun. When a comet approaches the Sun, the ices vaporize and the comet grows brighter, often developing a tail.

conjunction The alignment of two celestial bodies. When a planet is in conjunction with the Sun, it lies in the same direction and can't be seen.

constellation One of 88 regions of sky, usually encompassing a pattern of stars recognized since antiquity.

corona The outer layer of the Sun's atmosphere.

coronal mass ejection A large discharge of material from the Sun's corona that sweeps outward through the solar system.

crater A roughly circular depression, most often the result of an impact, found on objects throughout the solar system.

crescent A phase of the Moon or an inferior planet when less than half the disk appears lit.

dark adaptation The process by which the human eye adjusts to seeing dim objects under a dark sky.

day A single rotation of a planet or other solar system body on its axis.

declination A celestial coordinate equivalent to earthly latitude.

degree of arc An angle equal to 1⁄360 of a complete circle.

dew cap An extension to a telescope that prevents dew from condensing on the front lens.

diagonal A flat mirror used in a Newtonian telescope to reflect light from the primary to the eyepiece.

diamond ring During a total solar eclipse, the last glint of sunlight seen just before totality and the first glint seen after totality.

disk The visible extent of a celestial object. Only the Moon's and Sun's disk can be seen without a telescope.

Dobsonian mount A simple altazimuth mount capable of holding a large telescope.

double star Two stars that appear next to each other in the sky; typically, the two are in orbit around each other. The latter type is also called a binary star.

dust tail The part of a comet's tail composed of dust and glowing by reflecting sunlight.

eclipse The darkening of an object as it passes through the shadow cast by another object.

eclipsing binary A binary star in which the two members orbit each other in our line of sight, so that they periodically block each other from our view and dim accordingly.

ecliptic The apparent path of the Sun across our sky caused by Earth's orbital motion.

elliptical galaxy A galaxy with a spherical or elliptical shape and no trace of spiral structure.

elongation The angular distance between the Sun and a planet, or between a planet and one of its moons.

emission nebula A cloud of interstellar gas that glows by re-emitting radiation it absorbs from hot stars inside it.

equator The imaginary great circle on an object's surface halfway between the poles and perpendicular to its rotational axis.

equatorial mounting A telescope mounting in which one axis aligns with Earth's equator and the other with Earth's poles.

equinox The two points in an object's orbit of the Sun where the Sun shines down directly on the object's equator.

exit pupil The diameter of the light cone coming from the eyepiece of a telescope or binoculars.

eyepiece A lens or combination of lenses used to view an image created by the primary mirror or lens.

field of view The angular diameter of the area seen through an optical device.

filament A sinuous dark feature seen against the Sun's disk.

filter A device that blocks some wavelengths of light while transmitting others.

finderscope A low-power telescope attached to and aligned with a bigger scope to make it easier to locate objects.

fireball A particularly bright meteor, typically defined as being brighter than Venus.

First Quarter Moon The half-lit phase we see when the Moon has completed one-quarter of its orbit from one New Moon to the next.

focal length The distance between the primary lens or mirror and the focus in an optical device.

focal ratio The ratio between an optical system's focal length and the diameter of the primary lens or mirror.

full phase The phase of a Moon or planet when it appears completely lit.

galaxy A gravitationally bound system of up to a trillion or more stars, often with vast quantities of gas and dust. Galaxies are the fundamental building blocks of the universe.

Galilean moons The four biggest and brightest moons of Jupiter, discovered by Galileo when he first turned his telescope on Jupiter in 1610.

gas-giant planet A planet made mostly of hydrogen and helium gas. In our solar system, Jupiter, Saturn, Uranus, and Neptune are gas giants.

gibbous The phase of a Moon or planet when it appears more than half-lit but less than full.

globular cluster A tightly bound, spherical collection of hundreds of thousands of stars.

gravity An attractive force between all objects possessing mass.

Great Red Spot A huge storm system in Jupiter's atmosphere that has been raging for more than 300 years.

greatest elongation The maximum angular distance between the Sun and an inferior planet (Mercury or Venus) during an apparition.

horizon Where Earth and sky meet (apparent horizon), or the great circle located 90° from the zenith.

inferior conjunction The configuration when Mercury or Venus lies between the Sun and Earth.

inferior planet A planet with a smaller orbit than Earth's; only Mercury and Venus qualify.

interstellar medium The gas and dust that pervades the space between our galaxy's stars.

ion tail The part of a comet's tail composed of ionized gas that glows by re-emitting absorbed sunlight.

irregular galaxy A galaxy that appears disordered, with no prominent spiral or elliptical shape.

Kuiper Belt A disk-shaped region of the solar system beyond the orbit of Neptune. It likely contains millions of icy objects and is a storehouse of short-period comets.

Last Quarter Moon The half-lit phase we see when the Moon has completed three-quarters of its orbit from one New Moon to the next.

lens A transparent material, usually glass or plastic, designed with at least one curved surface so that light passing through will form an image.

light-year The distance light travels in one year. At 186,000 miles per second, this works out to approximately 5.9 trillion miles.

limb The edge of a celestial object's disk.

limiting magnitude The magnitude of the faintest stars visible with the eye or through an optical device.

Local Group The collection of a few dozen galaxies whose largest members are the Milky Way and Andromeda galaxies.

luminosity The total energy a star or other celestial object radiates into space.

lunar eclipse The passage of the Moon into Earth's shadow.

magnification The number of times larger an object appears when viewed through an optical device compared with its size seen with the naked eye.

magnitude A measure of the brightness of a celestial object. (See also absolute magnitude and apparent magnitude.)

mare One of the dark, lava-filled impact basins seen on the Moon.

Messier object One of 109 deep-sky objects cataloged by the 18th-century French observer Charles Messier. His catalog contains many of the sky's finest objects.

meteor The streak of light seen when a meteoroid enters Earth's atmosphere and air friction incinerates it. Meteors are often called "shooting stars."

meteor shower A display of meteors that arises when Earth passes through a stream of meteoroids. About eight major showers occur annually.

meteorite A meteoroid that survives its trip through the atmosphere and lands on Earth's surface.

meteoroid Bits of rock, most much smaller than boulder-sized, in orbit around the Sun. Meteoroids are invisible unless they encounter Earth.

Milky Way The band of light in the night sky that represents our galaxy's disk seen edge-on. Also, the name of our home galaxy.

mirror An optical component that reflects light. In a reflecting telescope, the primary mirror collects the light.

moon An object that orbits a planet. Earth's Moon is the brightest object in the night sky.

mounting A telescope's support structure.

multiple star A system of at least two stars in orbit around one another. If the system contains just two members, it's also called a binary star.

nautical twilight The level of illumination experienced when the Sun lies between 6° and 12° below the horizon.

nebula A cloud of interstellar gas and dust.

neutron star The crushed remains of a massive star that exploded as a supernova. A typical neutron star possesses more mass than the Sun in a sphere several miles across.

New General Catalogue (NGC) A catalog of more than 7,000 deep-sky objects, which includes most of the objects backyard observers target.

new phase The phase of a Moon or inferior planet when its lit half faces away from Earth and we can't see it.

Newtonian telescope The first type of reflecting telescope, invented by Isaac Newton, which uses a parabolic mirror to gather light and a flat secondary to redirect the light out the tube's side.

nova A star that experiences an explosion on its surface and temporarily brightens hundreds or thousands of times.

nucleus A comet's solid core of ice and rock. When a comet approaches the Sun, the ices vaporize and the nucleus gets hidden behind a veil of gas and dust.

objective The primary lens of a refracting telescope.

occultation The passage of an object with a large angular diameter in front of a smaller object, partially or totally blocking it from view.

Oort Cloud A vast spherical cloud, containing perhaps trillions of comets, that surrounds the Sun and stretches up to a light-year from our star.

open star cluster Also known as open cluster. A collection of hundreds or thousands of stars only loosely bound by gravity. Most will disperse in less than a billion years.

opposition A configuration in which a superior planet or smaller object lies opposite the Sun in our sky. A planet shines brightest at opposition and remains visible all night.

orbit The path of one object around a larger one.

peculiar galaxy A galaxy with an odd shape that doesn't fit within the normal classification scheme. Most peculiar galaxies seem to be distorted by a neighboring galaxy.

penumbra The lighter, outer part of a shadow or a sunspot. During an eclipse, the penumbra marks the region where the Sun is partially visible.

perigee The point where an object orbiting Earth (like the Moon) lies closest to our planet.

perihelion The point where an object orbiting the Sun lies closest to our star.

period The time interval between two successive occurrences of a regular event. For example, the time for an object to complete one revolution.

periodic comet A comet that returns regularly to the vicinity of the Sun. A short-period comet has a period of less than 200 years.

phase The illuminated fraction of the Moon's or other celestial object's disk.

photosphere The apparent surface of the Sun, from which the vast majority of its light radiates.

planet One of the eight large objects that circle the Sun. Astronomers have also discovered hundreds of planets in orbit around other stars.

planetary nebula A glowing cloud of gas gently ejected by an aging red-giant star.

primary mirror The main light-gathering component of a reflecting telescope.

prograde motion The dominant west-to-east motion exhibited by most solar system objects most of the time.

prominence A large tongue of gas arching above the Sun's limb. To see prominences, an observer needs a hydrogen-alpha filter or to witness a total solar eclipse.

pulsar A rapidly spinning neutron star that emits periodic bursts of radiation. The most famous pulsar lies at the heart of the Crab Nebula supernova remnant.

quadrature The configuration of a planet or other solar system object when its elongation from the Sun is 90°.

radiant The point from which all the meteors in a shower appear to radiate.

red giant A bloated, aging star with a relatively cool surface temperature.

reflecting telescope A telescope that uses a mirror to gather light.

reflection nebula An interstellar cloud that scatters and reflects starlight.

refracting telescope A telescope that uses a lens to gather light.

refraction The bending of light rays that occurs when they pass from one transparent medium to another.

resolution The amount of fine detail visible in an image.

retrograde motion The apparent east-to-west motion relative to the background stars of a planet or smaller solar system object.

revolution The motion of one body around another.

rich-field telescope A telescope that delivers a particularly large field of view.

right ascension A celestial coordinate equivalent to earthly longitude.

rotation The spinning of a body around its axis.

satellite A small object in orbit around a bigger one. Natural satellites are usually called moons.

Schmidt-Cassegrain telescope A compound telescope featuring a corrector plate at the front end and a spherical primary mirror. After the light reaches the primary mirror, it travels the same route as in a Cassegrain telescope.

seeing A measure of the steadiness of Earth's atmosphere, and thus how sharp celestial objects will appear.

separation The angular distance between two celestial bodies.

solar cycle An approximately 11-year cycle in which the Sun sports greater and fewer numbers of sunspots and other signs of activity.

solar eclipse The passage of Earth into the Moon's shadow.

solar system The Sun and all the objects that fall under its gravitational influence.

solar wind A stream of charged particles that flows away from the Sun.

solstice The two points in an object's orbit of the Sun where the Sun appears farthest north or south of the object's equator.

spiral galaxy A galaxy featuring a central bulge of mostly old stars surrounded by arms containing younger stars and the gas clouds that spawn them.

sporadic meteor A meteor not associated with any shower.

star A self-luminous sphere of gas.

Sun The star at the center of our solar system. The Sun holds more than 99 percent of the solar system's mass.

sunspot A dark, temporary, and relatively cool area in the Sun's photosphere.

superior conjunction The configuration when Mercury or Venus lies on the far side of the Sun from Earth.

superior planet A planet with a larger orbit than Earth's.

supernova An exploding star that for a few weeks can shine as brightly as all the other stars in a galaxy combined.

supernova remnant The gaseous remains of a massive star that exploded as a supernova.

surface brightness The apparent brightness of each small part of an extended object.

synchronous rotation The state of an orbiting object, like the Moon, when it takes the same length of time to rotate as it does to revolve. Such an object always keeps the same face toward its parent object.

tail A sometimes lengthy appendage to a comet consisting of gas and dust released by the nucleus. Solar radiation and the solar wind push this material away from the Sun.

telescope An optical device that gathers, focuses, and magnifies light.

terminator The line dividing day from night on an object's surface.

terrestrial planet A planet made mostly of rock and metal. In our solar system, Mercury, Venus, Earth, and Mars are terrestrial planets.

transit The passage of a small object in front of a larger one. The most noticeable such events occur when Mercury or Venus transits the Sun's disk, or the Galilean moons transit Jupiter.

transparency The clarity of the sky.

twilight The illumination present before sunrise or after sunset when the Sun lies less than 18° below the horizon.

umbra The dark, inner part of a shadow or a sunspot. During a solar eclipse, the umbral shadow marks the region where the Sun is totally eclipsed.

variable star A star whose light output varies, either because of intrinsic changes to the star or because another star periodically passes in front of it.

visual binary A double star that arises through a chance alignment of two objects at different distances.

waning The phase of the Moon or other solar system object as the illuminated part of its disk grows smaller.

waxing The phase of the Moon or other solar system object as the illuminated part of its disk grows larger.

white dwarf A star that has exhausted its nuclear fuel and been squeezed by gravity so its diameter is about the size of Earth.

year The time it takes an object to complete one revolution around the Sun.

zenith The point on the celestial sphere directly above an observer's head.

zodiac A band around the sky centered on the ecliptic and home to all the planets.

zodiacal light A faint illumination in the shape of a pyramid that appears after evening twilight ends and before morning twilight begins. It comes from sunlight reflecting off dust in the inner solar system.

The 88 Constellations

The 88 Constellations			
Name	***English Description***	***Genitive Case***	***Best Seen****
Andromeda	Chained Princess	Andromedae	Autumn
Antlia	Air Pump	Antliae	Winter
Apus	Bird of Paradise	Apodis	South
Aquarius	Water-bearer	Aquarii	Summer
Aquila	Eagle	Aquilae	Summer
Ara	Altar	Arae	Summer
Aries	Ram	Arietis	Autumn
Auriga	Charioteer	Aurigae	Winter
Boötes	Herdsman	Boötis	Spring
Caelum	Chisel	Caeli	Winter
Camelopardalis	Giraffe	Camelopardalis	Winter
Cancer	Crab	Cancri	Winter
Canes Venatici	Hunting Dogs	Canum Venaticorum	Spring
Canis Major	Big Dog	Canis Majoris	Winter
Canis Minor	Little Dog	Canis Minoris	Winter
Capricornus	Sea Goat	Capricorni	Summer
Carina	Keel	Carinae	Winter
Cassiopeia	Queen	Cassiopeiae	Autumn
Centaurus	Centaur	Centauri	Spring
Cepheus	King	Cephei	Autumn
Cetus	Whale	Ceti	Autumn
Chamaeleon	Chameleon	Chamaeleontis	South
Circinus	Compasses	Circini	South
Columba	Dove	Columbae	Winter
Coma Berenices	Berenice's Hair	Comae Berenices	Spring
Corona Australis	Southern Crown	Coronae Australis	Summer
Corona Borealis	Northern Crown	Coronae Borealis	Spring
Corvus	Crow	Corvi	Spring
Crater	Cup	Crateris	Spring
Crux	Cross	Crucis	South
Cygnus	Swan	Cygni	Summer

"Best seen" gives the season in which the constellation's center appears highest around midnight. Those constellations that lie too far south to see from mid-northern latitudes are labeled "South."

Name	English Description	Genitive Case	Best Seen
Delphinus	Dolphin	Delphini	Summer
Dorado	Goldfish	Doradus	South
Draco	Dragon	Draconis	Spring
Equuleus	Little Horse	Equulei	Summer
Eridanus	River	Eridani	Autumn
Fornax	Furnace	Fornacis	Autumn
Gemini	Twins	Geminorum	Winter
Grus	Crane	Gruis	Summer
Hercules	Hero	Herculis	Summer
Horologium	Clock	Horologii	South
Hydra	Water Snake	Hydrae	Spring
Hydrus	Little Water Snake	Hydri	South
Indus	Indian	Indi	Summer
Lacerta	Lizard	Lacertae	Summer
Leo	Lion	Leonis	Spring
Leo Minor	Little Lion	Leonis Minoris	Winter
Lepus	Hare	Leporis	Winter
Libra	Scales	Librae	Spring
Lupus	Wolf	Lupi	Spring
Lynx	Lynx	Lyncis	Winter
Lyra	Harp	Lyrae	Summer
Mensa	Table Mountain	Mensae	South
Microscopium	Microscope	Microscopii	Summer
Monoceros	Unicorn	Monocerotis	Winter
Musca	Fly	Muscae	South
Norma	Square	Normae	Spring
Octans	Octant	Octantis	South
Ophiuchus	Serpent-bearer	Ophiuchi	Summer
Orion	Hunter	Orionis	Winter
Pavo	Peacock	Pavonis	South
Pegasus	Winged Horse	Pegasi	Autumn

CONTINUED ON NEXT PAGE

The 88 Constellations (continued)

The 88 Constellations *(continued)*			
Name	***English Description***	***Genitive Case***	***Best Seen***
Perseus	Hero	Persei	Autumn
Phoenix	Phoenix	Phoenicis	Autumn
Pictor	Painter	Pictoris	Winter
Pisces	Fish	Piscium	Autumn
Piscis Austrinus	Southern Fish	Piscis Austrini	Summer
Puppis	Stern	Puppis	Winter
Pyxis	Compass	Pyxidis	Winter
Reticulum	Reticle	Reticuli	South
Sagitta	Arrow	Sagittae	Summer
Sagittarius	Archer	Sagittarii	Summer
Scorpius	Scorpion	Scorpii	Summer
Sculptor	Sculptor	Sculptoris	Autumn
Scutum	Shield	Scuti	Summer
Serpens	Serpent	Serpentis	Summer
Sextans	Sextant	Sextantis	Winter
Taurus	Bull	Tauri	Winter
Telescopium	Telescope	Telescopii	Summer
Triangulum	Triangle	Trianguli	Autumn
Triangulum Australe	Southern Triangle	Trianguli Australis	South
Tucana	Toucan	Tucanae	South
Ursa Major	Great Bear	Ursae Majoris	Spring
Ursa Minor	Little Bear	Ursae Minoris	All year
Vela	Sails	Velorum	Winter
Virgo	Maiden	Virginis	Spring
Volans	Flying Fish	Volantis	South
Vulpecula	Fox	Vulpeculae	Summer

The 25 Brightest Stars

The 25 Brightest Stars			
Star Name	***Constellation***	***Magnitude****	***Distance***
Sirius	Canis Major	–1.47	8.58
Canopus	Carina	–0.72	313.00
Alpha Centauri	Centaurus	–0.29	4.36
Arcturus	Boötes	–0.04	36.70
Vega	Lyra	0.03	25.30
Capella	Auriga	0.08	42.20
Rigel	Orion	0.12	773.00
Procyon	Canis Minor	0.34	11.40
Achernar	Eridanus	0.50	144.00
Betelgeuse	Orion	0.58	522.00
Beta Centauri	Centaurus	0.60	525.00
Altair	Aquila	0.77	16.80
Aldebaran	Taurus	0.85	65.10
Acrux	Crux	0.94	321.00
Spica	Virgo	1.04	262.00
Antares	Scorpius	1.09	604.00
Pollux	Gemini	1.15	37.60
Fomalhaut	Piscis Austrinus	1.16	25.10
Deneb	Cygnus	1.25	1,500.00
Mimosa	Crux	1.30	352.00
Regulus	Leo	1.35	77.50
Adhara	Canis Major	1.51	431.00
Castor	Gemini	1.59	51.50
Shaula	Scorpius	1.62	359.00
Gamma Crucis	Crux	1.63	87.90

**Magnitude means apparent visual magnitude and include all components in multiple-star systems; distances are in light-years.*

Telescope Resources

The following companies all manufacture and sell quality telescopes. Check out their Web sites to see what equipment they offer. These and other companies also sell accessories, like eyepieces and filters. Phil Harrington offers up-to-date information on astronomy manufacturers and dealers at his *Star Ware* Web site, www.philharrington.net.

Telescope Manufacturers

Apogee. Inc.
Phone: (800) 504-5897
Web site: www.apogeeinc.com

Astro-Physics, Inc.
Phone: (815) 282-1513
Web site: www.astro-physics.com

Celestron, LLC
Phone: (310) 328-9560
Web site: www.celestron.com

Coronado
Phone: (800) 626-3233
Web site: www.coronadofilters.com

D & G Optical
Phone: (717) 665-2076
Web site: www.dgoptical.com

Discovery Telescopes
Phone: (877) 523-4400
Web site: www.discovery telescope.com

Edmund Scientific Co.
Phone: (800) 728-6999
Web site: www.scientificsonline.com

Jim's Mobile, Inc. (JMI)
Phone: (800) 247-0304
Web site: www.jmitelescopes.com

Konus USA
Phone: (305) 592-5500
Web site: www.konus.com

Mag One Instruments
Phone: (920) 849-9151
Web site: www.mag1 instruments.com

Meade Instruments Corp.
Phone: (800) 626-3233
Web site: www.meade.com

Obsession Telescopes
Phone: (920) 648-2328
Web site: www.obsession telescopes.com

Orion Telescopes and Binoculars
Phone: (800) 447-1001
Web site: www.telescope.com

Pacific Telescope Corp.
Phone: (604) 241-7027
Web site: www.skywatcher telescope.com

Parks Optical Inc.
Phone: (805) 522-6722
Web site: www.parksoptical.com

Questar Corporation
Phone: (215) 862-5277
Web site: www.questar-corp.com

RC Optical Systems
Phone: (928) 526-5380
Web site: www.rcopticalsystems.com

Sky Instruments
Phone: (800) 648-4188
Web site: www.antaresoptical.com

Starmaster Telescopes
Phone: (620) 638-4743
Web site: www.starmaster telescopes.com

Stellarvue Telescopes
Phone: (530) 823-7796
Web site: www.stellarvue.com

Takahashi Seisakusho Ltd.
U.S. Importer: Land, Sea and Sky
Phone: (713) 529-3551
Web site: www.takahashi america.com

Telescope Engineering Company
Phone: (303) 273-9322
Web site: www.telescopengineering.com

Tele Vue Optics, Inc.
Phone: (845) 469-4551
Web site: www.televue.com

TMB Optical LLC
Phone: (216) 524-1107
Web site: www.tmboptical.com

TScopes
Web site: www.tscopes.com

Vixen Company, Ltd.
Phone: (845) 469-8660
Web site: www.vixenamerica.com

Art Credits

CHAPTER ONE

Pages 4–5, Illustrations by Richard Talcott

Page 7, photos, courtesy of Frederick A. Ringwald

Page 8, illustration, courtesy of Roen Kelly; photo, courtesy of Mike Salway

Pages 9–10, illustrations, courtesy of Roen Kelly

Page 11, illustration by Richard Talcott

Page 12, photos, courtesy of Evelyn Talcott

Page 13, top photo, courtesy of SOHO (ESA and NASA); middle photo, courtesy of Mike Salway; bottom photo, courtesy of NASA/ESA/AURA/Caltech

Page 14, illustrations by Richard Talcott

Page 15, illustration and photo by Richard Talcott

Page 16, photos, courtesy of NASA

Page 17, photo, courtesy of Bill and Sally Fletcher

Page 18, top photo, courtesy of Mike Simmons; bottom photo, courtesy of Marvin Nauman

Page 19, photo, courtesy of Mike Salway

CHAPTER TWO

Page 22, top photo, courtesy of *Astronomy* magazine, Kalmbach Publishing Co.; middle and bottom photos, courtesy of Evelyn Talcott

Page 23, photos, courtesy of Evelyn Talcott

Page 24, photos, courtesy of *Astronomy* magazine, Kalmbach Publishing Co.

Page 25, illustrations, courtesy of *Astronomy* magazine, Kalmbach Publishing Co.

Pages 26–27, photos, courtesy of *Astronomy* magazine, Kalmbach Publishing Co.

Page 28, top photo, courtesy of Michael E. Bakich; bottom photo, courtesy of *Astronomy* magazine, Kalmbach Publishing Co.

Page 29, photos, courtesy of *Astronomy* magazine, Kalmbach Publishing Co.

CHAPTER THREE

Pages 32–33, photos, courtesy of Alister Ling

Page 34, top photo, courtesy of Marvin Nauman; bottom photo, courtesy of Alister Ling

Page 35, top photo, courtesy of Alister Ling; bottom photo, courtesy of Marvin Nauman

Page 36, photos by Richard Talcott

Page 37, photo, courtesy of Marvin Nauman

Page 38, top photo, courtesy of Alister Ling; bottom photo by Richard Talcott

Page 39, photo, courtesy of Mike Salway

Pages 40–41, photos, courtesy of Rob Ratkowski

Page 42, top photo, courtesy of SOHA (ESA and NASA); bottom photo, courtesy of Marvin Nauman

Page 43, photo, courtesy of Marvin Nauman

Page 44, top photo, courtesy of Alister Ling; middle and bottom photos, courtesy of Marvin Nauman

Page 45, photo, courtesy of Marvin Nauman

Page 46, photo by Richard Talcott

Page 47, top photo, courtesy of Alister Ling; bottom photo, courtesy of Oshin D. Zakarian

Page 48, illustrations, courtesy of *Astronomy* magazine/Kalmbach Publishing Co.

Page 49, photo, courtesy of Marvin Nauman

Page 50, top photo by Richard Talcott; bottom photo, courtesy of Alister Ling

Page 51, photo, courtesy of Rob Ratkowski

Page 52, top photo, courtesy of NASA/JPL-Caltech; bottom photo, courtesy of Mike Salway

Page 53, photo, courtesy of Rob Ratkowski

CHAPTER FOUR

Pages 57–58, illustrations by Richard Talcott

Page 59, photo, courtesy of Bill and Sally Fletcher

Page 60, top photo, courtesy of Bill and Sally Fletcher; middle left photo, courtesy of NOAO/AURA/NSF; middle right photo, courtesy of NASA/ESA/AURU/Caltech; bottom photo, courtesy of Bill Schoening/NOAO/AURA/NSF

Page 61, photo, courtesy of NASA/ESA/J. Hester and A. Loll (Arizona State University)

Page 62, illustration by Richard Talcott

Page 63, photo, courtesy of Bill and Sally Fletcher

Page 64, top photo, courtesy of Robert Gendler; middle photo, courtesy of NOAO/AURA/NSF; bottom photo, courtesy of Jim Rada/Adam Block/NOAO/AURA/NSF

Page 65, photo, courtesy of NASA/ESA/M. Robberto (STScI/ESA)/The HST Orion Treasury Project

Page 66, illustration by Richard Talcott

Page 67, photo, courtesy of Bill and Sally Fletcher

Page 68, top photo, courtesy of Frederick A. Ringwald; middle left and right photos, courtesy of N.A. Sharp/NOAO/AURA/NSF; bottom photo, courtesy of Peter and Suzie Erickson/Adam Block/NOAO/AURA/NSF

Page 69, photo, courtesy of NASA/Andy Fruchter and the ERO Team

CHAPTER FIVE

Pages 73–74, illustrations by Richard Talcott

Page 75, photo, courtesy of Bill and Sally Fletcher

Pages 76–77, photos, courtesy of Robert Gendler

Page 78, illustration by Richard Talcott

Page 79, photo, courtesy of Bill and Sally Fletcher

Page 80, top photo, courtesy of Bill and Sally Fletcher; middle and bottom photos courtesy of Robert Gendler

Page 81, photo, courtesy of NASA/ESA

Page 82, illustration by Richard Talcott

Page 83, photo, courtesy of Bill and Sally Fletcher

Page 84, top photo, courtesy of Adam Block/NOAO/AURA/NSF; middle photo, courtesy of Robert Gendler; bottom photo, courtesy of NASA/ESA/The Hubble Heritage Team (STScI/AURA)

Page 85, photo, courtesy of Robert Gendler

CHAPTER SIX

Pages 89–90, illustrations by Richard Talcott

Page 91, photo, courtesy of Bill and Sally Fletcher

Page 92, top photo, courtesy of Robert Gendler; middle photo, courtesy of NASA/ESA/A. Zijlstra (UMIST); bottom photo, courtesy of Robert Gendler

Page 93, photo, courtesy of NASA/ESA/The Hubble Heritage Team (STScI/AURA)

Page 94, illustration by Richard Talcott

Page 95, photo, courtesy of Bill and Sally Fletcher

Page 96, top photo, courtesy of Fred Calvert/Adam Block/NOAO/AURA/NSF; middle photo, courtesy of NASA/ESA/H. Ford (JHU) and the ACS Science Team; bottom photo, courtesy of Robert Gendler

Page 97, photo, courtesy of Robert Gendler

Page 98, illustration by Richard Talcott

Page 99, photo, courtesy of Bill and Sally Fletcher

Page 100, top photo, courtesy of Robert Gendler; middle photo, courtesy of Bruce Balick (University of Washington), et al. and NASA; bottom photo, courtesy of Robert Gendler

Page 101, photo, courtesy of Robert Gendler

CHAPTER SEVEN

Pages 105–106, illustrations by Richard Talcott

Page 107, photo, courtesy of Bill and Sally Fletcher

Page 108, top photo, courtesy of Bill and Sally Fletcher; middle photo, courtesy of Robert Gendler; bottom photo, courtesy of Steve and Paul Mandel/Adam Block/NOAO/AURA/NSF

Page 109, photo, courtesy of Robert Gendler

Page 110, illustration by Richard Talcott

Page 111, photo courtesy of Bill and Sally Fletcher

Pages 112–113, photos, courtesy of Robert Gendler

Page 114, illustration by Richard Talcott

Page 115, photo, courtesy of Bill and Sally Fletcher

Page 116, top photo, courtesy of Dale Cupp/Flynn Haase/NOAO/AURA/NSF; middle photo, courtesy of Adam Block/NOAO/AURA/NSF; bottom photo, courtesy of Robert Gendler

Page 117, photo, courtesy of Robert Gendler

CHAPTER EIGHT

Page 120, illustration by Richard Talcott; bottom photo, courtesy of Marvin Nauman

Page 121, photos, courtesy of Michael E. Bakich

Page 122, top photo, courtesy of Royal Swedish Academy of Science; bottom illustration by Richard Talcott

Page 123, photo, courtesy of SOHO (ESA and NASA)

Pages 124–125, photos, courtesy of Jack Newton

Page 126, illustration by Richard Talcott

Page 127, photo, courtesy of Bill and Sally Fletcher

Page 128, photos, courtesy of Eric Douglas and the Lunar and Planetary Institute

Page 129, photo, courtesy of Adam Block/NOAO/AURA/NSF

Page 130, photos, courtesy of Eric Douglas and the Lunar and Planetary Institute

Page 131, photo, courtesy of Anthony Ayiomamitis

Page 132, photos, courtesy of Eric Douglas and the Lunar and Planetary Institute

Page 133, photo, courtesy of Anthony Ayiomamitis

Page 134, photos, courtesy of Eric Douglas and the Lunar and Planetary Institute

Page 135, photo, courtesy of Robert Gendler

Page 136, top photo, courtesy of Alister Ling; bottom photo, courtesy of Luca Vanzella

Page 137, photo, courtesy of Anthony Ayiomamitis

Page 138, top illustration by Richard Talcott; bottom photo, courtesy of Alister Ling

Page 139, photo, courtesy of Anthony Ayiomamitis

Page 140, top and middle photos, courtesy of U.S. Navy/Joshua Valcarcel; bottom photo, courtesy of Mike Salway

Page 141, photo, courtesy of Bill and Sally Fletcher

Page 142, illustrations, courtesy of Richard Talcott

Page 143, photo, courtesy of Mike Simmons

Page 144, top photo, courtesy of Anthony Ayiomamitis; middle photo by Richard Talcott; bottom photo, courtesy of Martin Ratcliffe

Page 145, photo, courtesy of Anthony Ayiomamitis

Page 146, top photo, courtesy of James Manning; middle and bottom illustrations, courtesy of *Astronomy* magazine, Kalmbach Publishing Co.

Page 147, photo, courtesy of Anthony Ayiomamitis

CHAPTER NINE

Page 150, illustrations by Richard Talcott

Page 151, top photo, courtesy of Mike Salway; bottom photo, courtesy of NASA/JHUAPL/CIW

Page 152, top and middle photos, courtesy of Mike Salway; bottom photo, courtesy of Francis Reddy

Page 153, photos, courtesy of NASA/JPL

Page 154, illustrations by Richard Talcott

Page 155, photos, courtesy of Mike Salway

Page 156, photos, courtesy of NASA/J. Bell (Cornell University) and M. Wolff (SSI)

Page 157, photo, courtesy of NASA/JPL/Cornell University

Page 158, top photo, courtesy of NASA/JPL/University of Arizona; bottom photo, courtesy of NASA/ESA/M. Wong and I. de Pater (University of California, Berkeley)

Pages 159–160, photos, courtesy of Mike Salway

Page 161, photo, courtesy of NASA/JPL/SSI

Page 162, photo, courtesy of Mike Salway

Page 163, photos, courtesy of NASA/JPL/SSI

Pages 164–165, photos, courtesy of NASA/ESA/The Hubble Heritage Team (STScI/AURA)

Page 166, top photo, courtesy of Mike Salway; bottom photo, courtesy of Erich Karkoschka (University of Arizona)/NASA

Page 167, top photo, courtesy of NAS/L. Sromovsky and P. Fry (University of Wisconsin); bottom photo, courtesy of NASA/JPL

Page 168, top photo, courtesy of NASA/ESA/R. Albrecht (ESO-STECF); bottom illustration, courtesy of NASA/Ben Zellner (Georgia Southern University)/Peter Thomas (Cornell University)

Page 169, top photo,courtesy of NASA/JPL; bottom photo, courtesy of NASA/JPL/JHUAPL

Page 170, top photo, courtesy of NASA/JPL/UMD; bottom photo, courtesy of NASA/ESA/The Hubble Heritage Team (STScI/AURA)

Page 171, photo, courtesy of Bill and Sally Fletcher

Page 172, top photo, courtesy of NASA/H. Weaver and T. Smith (STScI); bottom photo, courtesy of NASA/R. Evans, J. Trauger, H. Hammel, and the HST Comet Science Team

Page 173, photo, courtesy of Mike Salway

CHAPTER TEN

Page 176, photos, courtesy of Bill and Sally Fletcher

Page 177, top photo and bottom right photo, courtesy of Frederick A. Ringwald; bottom left illustration by Richard Talcott

Pages 178–179, illustrations by Richard Talcott

Page 180, top photo, courtesy of NOAO/AURA/NSF; bottom photo, courtesy of Robert Gendler

Pages 181–185, photos, courtesy of Robert Gendler

Page 186, top photo courtesy of Robert Gendler; bottom photo, courtesy of T.A. Rector (NOAO/AURA/NSF) and the Hubble Heritage Team (STScI/AURA/NASA)

Page 187, photo, courtesy of Robert Gendler

Page 188, top photo, courtesy of Robert Gendler; bottom photo, courtesy of NASA/ESA/HEIC/The Hubble Heritage Team (STScI/AURA)

Page 189, photo, courtesy of NASA/ESA/The Hubble Heritage Team (STScI/AURA)

Page 190, top photo, courtesy of NASA/ESA/A. Filippenko (UC, Berkeley)/P. Challis (Harvard-Smithsonian CfA)/et al; bottom photo, courtesy of NASA/ESA/N. Smith (UC, Berkeley)/NOAO/AURA/NSF

Page 191, photo, courtesy of NASA/ESA/The Hubble Heritage Team (STScI/AURA)

Page 192, top photo, courtesy of Bruce Hugo and Leslie Gaul/Adam Block/NOAO/AURA/NSF; bottom photo, courtesy of NASA/ESA/The Hubble Heritage Team (STScI/AURA)

Page 193, photo, courtesy of NASA/ESA/S. Beckwith (STScI)/The Hubble Heritage Team (STScI/AURA)

Page 194, photos, courtesy of NASA/ESA/The Hubble Heritage Team (STScI/AURA)

Pages 195–196, photos, courtesy of Robert Gendler

Page 197, photo, courtesy of NASA/ESA/The Hubble Heritage Team (STScI/AURA)

Page 198, top photo, courtesy of NASA/J. English (University of Manitoba)/S. Hunsberger, S. Zonak, J. Charlton, and S. Gallagher (PSU)/L. Frattare (STScI); bottom photo, courtesy of Robert Gendler

Page 199, photo, courtesy of Robert Gendler

CHAPTER ELEVEN

Page 202, top photo and bottom sketch, courtesy of Martin Ratcliffe

Pages 203–205, sketches, courtesy of David J. Eicher

Page 207, photo, courtesy of Martin Ratcliffe

Page 208, photo, courtesy of Alister Ling

Page 209, photo, courtesy of Mike Salway

Page 210, photo, courtesy of Mike Simmons

Page 211, photo, courtesy of Mike Salway

Page 212, photo, courtesy of Anthony Ayiomamitis

Page 213, photo, courtesy of Alister Ling

Page 214, photo, courtesy of Marvin Nauman

Page 215, photo, courtesy of Anthony Ayiomamitis

Index